Technical Report Writing

Technical Report Writing

SECOND EDITION

JAMES W. SOUTHER
MYRON L. WHITE
University of Washington
Seattle, Washington

A Wiley-Interscience Publication

JOHN WILEY & SONS New York • London • Sydney • Toronto

Library of Congress Cataloging in Publication Data

Souther, James W
Technical report writing.

"A Wiley-Interscience publication."
Bibliography: p.
Includes index

1. Technical writing. 2. Report writing.
I. White, Myron Lester, 1918- joint author.
II. Title.
T11.S65 1977 808'.066'6 76-51461
ISBN 0-471-81412-1

Printed in the United States of America

10 9 8 7 6 5 4 3 2

Preface

Considerable rewriting and a substantial amount of new material went into the preparation of this second edition of *Technical Report Writing*. Those parts of the book most affected by changes and additions include the impact of the audience on writing, problems of organization, the use of illustrations and layout, problems of writing and revising, and problems of tone and style.

The design approach to writing, on which the original book was based, still serves as the foundation for this second edition. It, too, emphasizes the writing process, stressing those factors writers must consider and the steps of the process they must follow in producing an effective piece of writing. Indeed, the fact that it is process-based causes the book to differ most from others in the field. Strangely enough, this problem-solving approach to writing is almost as unique today as it was when the first edition appeared 20 years ago.

In writing this book we have considered as our primary audience the practicing professional, the person who knows the difficulty of communicating effectively and who seeks helpful ideas and suggestions. Needless to say, the book should also be useful to students who are preparing themselves for the world of work. In this second edition our goal has been to provide a book that would:

1. *Place emphasis on the writing process* which is common to all kinds of writing, whether the writer faces the task of writing a letter, a proposal, an impact statement, an article for publication, or documenting report. The product that the writer turns out varies from one writing situation to another. Only the process remains constant. Understanding the writing process is the most valid approach to the solution of writing problems not only today but also rapidly changing future.
2. *Provide an overall view of writing* that will allow the writer to fit together the many individual tasks that make up the process. Each task in the writing process is not independent, but is related to the others and to the sequence in which the writer takes them up. The impact of these interrelationships is as important to the quality of the finished product as the successful accomplishment of each individual task.
3. *Provide numerous practical, realistic, work-proved suggestions* that will help to make writing easier and more effective.

The widespread acceptance and endorsement of this design approach to writing by the many professional engineers and scientists in our short courses for industry and government testify to its validity, value, and workability. In fact, the acceptance of this approach by these engineers and scientists, as well as by the students in our university classes, has provided our motivation in offering this second and expanded edition.

The organization of the book, like that of the first

edition, tends to be process-based. The first two chapters are introductory. They establish the nature and characteristics of technical and scientific writing and describe the design approach to writing. Following this introductory overview, each of the next four chapters presents a more detailed view of one of the stages in the writing process. An additional chapter takes up the writing of abstracts and summaries. Then the brief epilogue sets the writing process in a realistic framework. It suggests that decisions about a report's level of quality must be brought into some kind of realistic balance with the amount of time one can spend in attaining that quality.

For convenience, we have presented the various stages of the writing process in separate chapters, although in practice these stages are not so clear-cut as this division would make them appear. In writing there is a kind of spiral development, and one stage often overlaps with another. Material that other books present in isolated subject matter compartments is in this book integrated into the writing process. Consequently, such matters as report form, the use of illustrations, and organizing content get discussed at the point in the process at which the writer comes face to face with them. Thus it will be helpful to readers if they become familiar with the entire book before attempting to use it as a reference. Familiarity with the stages of the writing process will assist them materially in locating the discussion of a particular topic or problem.

As other authors, we, too, are indebted to many people. Both of us gained more than we can ever express as students under the late Professor Porter G. Perrin. To our colleagues, Stuart W. Chapman, Mary B. Coney, Eugene C. Elliott, and Louis P. Trimble of the University of Washington, we owe much, for they have listened to our problems and have made many helpful suggestions. We are also indebted to many of our friends and colleagues in the Society for Technical Communication, who have encouraged us, listened to us, at times disagreed with us, but always supported us in our efforts to produce this second edition. Our students, too, both the professionals in government and industry and the students in our university classes, have contributed important insights into the problems of scientific and technical writing. Finally, we owe a special debt to Grace Blough and Marcella Hook, who not only put up with us during the many versions of the manuscript but also made valuable suggestions in the preparation and writing of this edition.

To these, and many more, the book owes a large part of its value, for each has contributed to the blending of writing theory and philosophy on the one hand with the practical dimensions of writing on the other. Whatever weaknesses it may have remain, of course, our responsibility.

JAMES W. SOUTHER
MYRON L. WHITE

Seattle, Washington
December 1976

Contents

CHAPTER 1 **Introduction: Scientific and Technical Writing, 1**

WHAT DISTINGUISHES SCIENTIFIC AND TECHNICAL WRITING? 1
- Written by Assignment, 2
- Two Problems to Solve, 2
  - The Technical Problem, 2
  - The Communication Problem, 3
- A Basis for Decision Making, 3

WHY IS SCIENTIFIC AND TECHNICAL WRITING DIFFICULT? 4

WHAT DOES THIS BOOK OFFER? 5

CHAPTER 2 **The Design Approach to Technical Writing, 6**

THE WRITING PROCESS, 6
- Analyzing the Communication Problem, 6
- Investigating the Alternatives, 7
- Designing the Report, 8
- Applying the Design, 8

IMPORTANT FEATURES OF THE WRITING PROCESS, 8

CHAPTER 3 **Analyzing the Communication Problem, 10**

PURPOSE, 10
- How Reports are Affected by Purpose, 10
- Classification of Reports by Purpose, 13
  - To Inform, 13
  - To Initiate Action, 13
  - To Coordinate Projects, 14
  - To Recommend, 14
  - To Provide a Record, 14

USE, 15
- Organization, 15
- Identifying Content, 16
- Choice of Form, 16
- Physical Characteristics, 16

AUDIENCE, 17
- Informational Needs, 17
  - Technical Management, 17
  - Technical Staff, 20
- Writing for the Reader, 21
  - What is Wanted? 21
  - What Else is Needed? 22
  - The Reader's Background, 23

CHAPTER 4 **Investigating the Alternatives, 25**

DEVELOPING CRITERIA, 25

IDENTIFYING CONTENT, 26
Defining Content, 26
Gathering Content, 29

SEARCHING FOR ALTERNATIVES, 30

SELECTING THE PATTERN OF PRESENTATION, 31

CHAPTER 5 **Designing the Report, 34**

SELECTING AND ORGANIZING CONTENT, 34
Constructing the Outline, 35
Selecting Material, 36
Organizing Material, 37
Establishing Relationships, 38
Establishing Order, 39
Revealing Organization, 40

PROVIDING ILLUSTRATIVE SUPPORT, 42
Two Functions in Reporting, 42
When to Use, 43
Designing and Locating Illustrations, 45

DETERMINING FORM AND LAYOUT, 49
Choosing Report Form, 49
Correspondence Forms, 50
Report Forms, 50
Choosing Internal Layout, 52
Page Layout, 55

CHAPTER 6 **Applying the Design, 59**

PROBLEMS OF GETTING STARTED, 59

NATURE OF THE WRITING ACTIVITY, 60
Writing the First Draft, 61
Revising, 62
Preparing Final Copy, 62

PROBLEMS OF TONE AND STYLE, 63
Tone, 63
Impersonal Tone, 64
Negative Tone, 66
Inflated Tone, 69
Style, 69
Paragraphs, 70
Listing Ideas, 70
Transitions, 71
Sentences, 72
Faulty Structure, 73
Wordiness, 74
Listing Details, 76
Words, 76
Inaccurate Words, 77
Unnecessarily "Big Words," 77
Stacking Modifiers, 78

CHAPTER 7 **Abstracts and Summaries, 82**

ABSTRACTS, 82

SUMMARIES, 83

CHAPTER 8 **Epilogue, 85**

Selected Bibliography, 87

Subject Index, 90

Technical Report Writing

CHAPTER 1

Introduction: Scientific and Technical Writing

Scientific and technical writing has assumed a role of major importance in the working world today. Since World War II it has exploded into unexpected dimensions, and Alvin Toffler has succinctly described the result in *Future Shock*. Toffler points out that although the rate at which human beings have been storing up knowledge about themselves and their universe has been spiraling "upward for a thousand years," the rate of increase is accelerating sharply today. For example, Toffler notes that "the number of scientific journals and articles is doubling, like industrial production in the advanced countries, about every 15 years." He also indicates the immense volume of scientific and technical writing by pointing out that "today the United States government alone generates 100,000 reports each year, plus 450,000 articles, books and papers. On a worldwide basis, scientific and technical literature mounts at a rate of some 60,000,000 pages a year."[1]

Not only engineers and scientists have felt the impact of this tremendous growth in communication of new knowledge. Everyone who lives in a highly technological society has felt it. In fact, the rising demand that citizens have a voice in the technological decisions affecting their lives has resulted from the impact scientific and technical writing is having on them. At the same time, their desire to participate in such decisions is creating new readers who wish to be informed about the advances in science and technology.

[1]Alvin E. Toffler, *Future Shock*, Random House, New York, © 1970, pp. 30–31.

This new demand is posing new problems for persons engaged in scientific and technical writing. It also emphasizes, however, that the basic purpose of such writing is, like that of all informative writing, to increase the knowledge of its readers. Most of the letters, reports, papers, articles, manuals, and books anyone writes are intended to inform, to educate, or to influence someone else. Scientific and technical writing is not exceptional in this respect. All the papers, reports, and pages of the scientific literature Toffler cites were written with the aim of telling someone about something.

WHAT DISTINGUISHES SCIENTIFIC AND TECHNICAL WRITING?

What, then, are the distinctive characteristics of scientific and technical writing? How does it differ from other kinds of informative writing? The most obvious difference is its content, which springs from science and technology. Frequently, as a result, it is mathematical, and it always requires a close attention to detail and accuracy. In addition, scientific and technical writing has other characteristics that are not so apparent, and these greatly influence what the writers must do and the writing they must produce.

Written By Assignment

Most scientific and technical writing, whether in business, industry, or government, results from assigned work. Writing grows out of work, and writers rarely find themselves able to write about subjects they choose. Frequently, they are asked to make judgments about their work, to offer conclusions or recommendations. Nevertheless, their writing centers on the work they have been assigned. Also, very often they must write according to specifications laid down by others—a customer, a government agency, or even their own organization. Whatever the source, the writing specifications are another constraint within which writers must cope. Thus writing by assignment and writing to "spec" are common features of scientific and technical writing.

Two Problems to Solve

Interestingly enough, the work scientists or engineers are assigned has more dimensions than they may be aware of. Actually, they are handed two very different problems when they are given a technical assignment and then asked to report on it. First, they must do the assigned technical work, make the study, gather the information, determine what it means. Next, they must report on that work to others.

When they shift from one task to the other, from the *doing* to the *reporting,* they too often fail to recognize that they face a new problem, a communication problem. Yet distinguishing the communication problem from the technical problem is critical. Those who fuse the two problems, thinking that there is only one, too frequently present their material in the same chronological order in which they did their technical work. Consequently, they can fail to communicate effectively.

Each of these problems requires its own solution. In solving a technical problem, engineers and scientists frequently work inductively, from detail to generalization. In solving a communication problem, however, they would do well to consider presenting their material deductively—inverting the order of their technical work—presenting the generalizations first and then providing the supporting detail.

The Technical Problem

In solving a technical problem, writers are "looking for answers." In seeking the solution to this problem, they must first determine their objectives and define their approach. Then, reducing the process to its simplest terms, they set out looking for answers by completing the following three steps:

1. *Conduct an extensive and detailed investigation of the problem.* This step is a process of gathering material and seeking out data through literature search, experimentation, consultation, and so on—bringing together data and the results relating to the specific problem.
2. *Examine and evaluate the material that has been gathered.* This step consists of a highly detailed examination of the data and results and identifying meaningful and important relationships. In the examination and evaluation, some of the material is discarded either as unimportant or perhaps unrelated.
3. *Make professional judgments about the material and the ideas growing out of it.* In this step the answer is defined, and conclusions, suggestions, or recommendations are formulated.

Solving the assigned technical problem, then, is really a process of gathering a great amount of data, and, with each successive analytical step, extracting the important and the significant out of the work that has gone before. As a result, less material is handled at each step. (See Figure 1, p. 3.)

Finding the solution to a technical problem is often a long and tedious process, one involving tremendous amounts of detail and painstaking attention. However, writers should recognize that all the information they examine in the process is not necessary to the solution of the communication problem. They must, by the very nature of the process of finding answers, examine minutely a mass of details to establish the relevance and significance of each one. Obviously, much of what they examine will be related and useful. Just as obviously, some of the material they gather and examine will not be useful.

The decision to include a specific detail in a report should depend on its relevance and its importance to the objectives of the work being reported, not on how much "sweat and blood" went into its gathering. Otherwise, writers can fail to distinguish between the "order of doing" and the "order of reporting." In this situation, steering what they consider the safe course, they are likely to include everything they have gathered and, in so doing, smother the significant information they want to communicate.

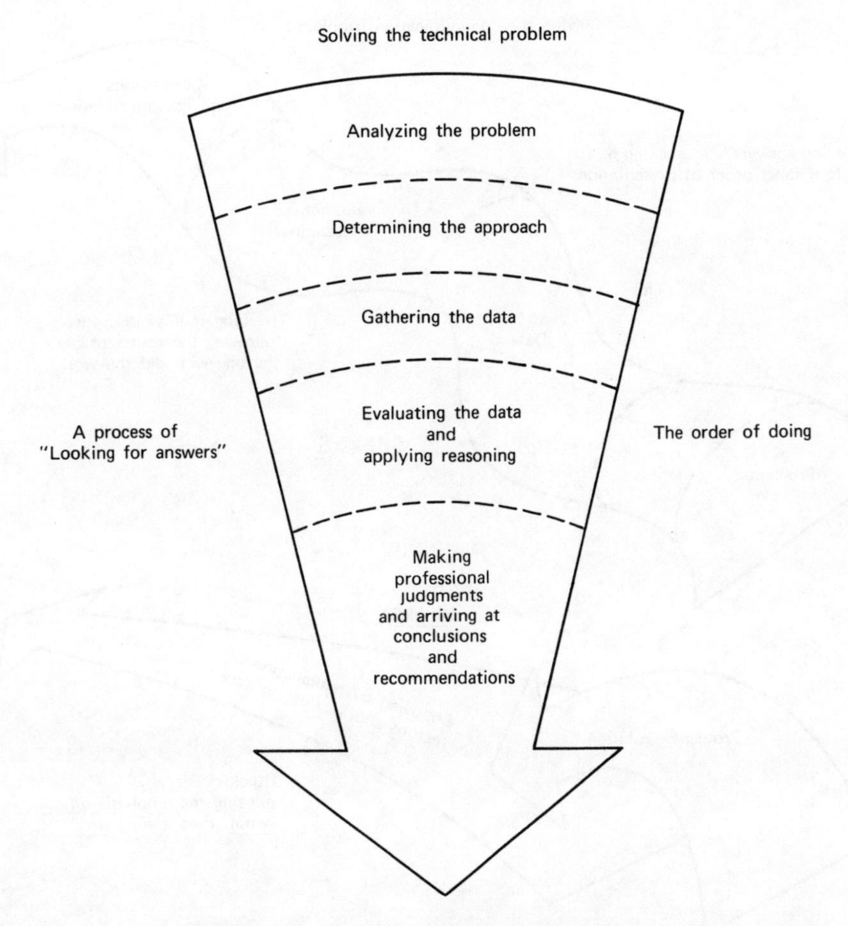

Figure 1. Process for solving technical problems.

The Communication Problem

The nature of the communication problem exerts its own influence on the reporting of the technical work. The purpose of the report, the use to be made of it, and the informational needs and background of the readers substantially influence the choice of content and its organization and presentation. To solve the communication problem, writers need to present their information in an order that is functional for their readers, and most readers are interested in the conclusions and the recommendations that grew out of the assigned work. The details of the investigation frequently interest the reader only as they relate to and support the generalizations that grow out of the study.

Distinguishing between "finding answers" and "telling answers" is extremely important to effective communication. As anyone who reads many reports is quick to point out, all too often reports are structured on the "order of doing," on the order used in carrying out the investigation. This chronological pattern forces the reader to follow the step-by-step procedure of the study, including, of course, the false leads, the dead ends, and the irrelevant detail. Emphasis is placed on the "long trail" and the minute detail instead of what grew out of the work. (See Figure 2, p. 4.)

Frequently, and for managers almost always, the reader is served better if writers invert the order of presenting their material, reversing the order of the steps they used in solving their technical problems.

A Basis for Decision Making

Another characteristic of scientific and technical writing is that it frequently serves as the basis for decision making. A report to a supervisor may well provide the information needed to make a decision on the deployment of manpower. A letter to a customer may well influence a decision to buy a piece of equipment. A report to a government body may well influence its

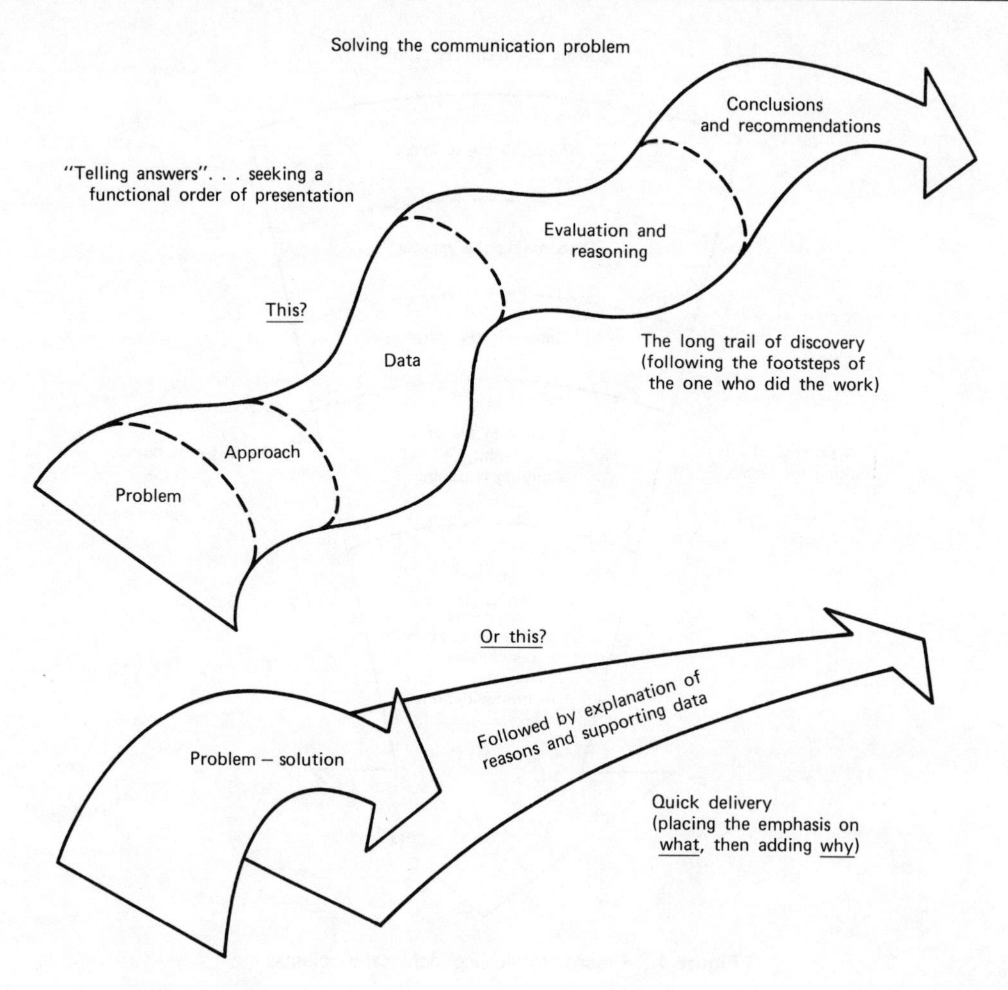

Figure 2. Solving communication problems requires that the author seek a functional order for presenting the report's content.

decision concerning the acceptability of a major project. An article informing the general public of the impact of a technological development proposed for an area may well influence public support for that development.

In each of these examples, the presentation of scientific and technological material serves as a basis for decision making. Frequently, engineers and scientists need to tailor their writing to the decision-making processes of individuals and decision-making groups. In each case they need to identify the kinds of information that will be essential to making a given decision. Then they must be sure to include as much of this information as they have available to them. They should organize their content so that the essential facts and ideas stand out for the reader, and they need to interpret the significance of these facts and ideas for the decision-making process. Responding to the needs of the decision maker is particularly important when one writes reports to management.

WHY IS SCIENTIFIC AND TECHNICAL WRITING DIFFICULT?

If it is to be good, all writing is difficult, because of the many decisions writers are forced to make at each stage of the writing process. They must make decisions very early, for example, about the nature of their audience, about their purpose in writing, and about the sources of information available to them, Later, they must determine the answers to other questions. What content should be included? What should be omitted? How should the content be organized? Finally, when they actually get down to drafting and revising a report, writers face another set of decisions. Is this the right word? Is that phrasing of an idea accurate? Is the relationship between these two ideas clear? Will the readers move without confusion from one section of the report to the next?

Unfortunately, many writers proceed on the basis of unconscious or unexamined assumptions when they prepare a report, rather than on the basis of conscious decisions related to their particular communication problem. Many others simply do not have sufficient knowledge about a given writing situation to provide them with a sound basis for the decisions they should make. In these circumstances, writing becomes more than difficult; it becomes downright frustrating. What writers need, of course, is an approach to writing that can provide them with the information they need, or at least make clear what their informational needs are.

WHAT DOES THIS BOOK OFFER?

The approach offered in this book is one that many engineers and scientists have used successfully. It insists that writing is not a single task, but rather a process or a series of related tasks by means of which writers can approach a solution to each writing problem they face. It also insists that engineers and scientists apply the problem-solving or design techniques they know so well to the solution of their communication problems.

In presenting a design approach to writing, this book provides an overview of the writing process as a whole; at the same time, it clearly defines each stage in that process. It suggests where writers should start and what steps they should follow thereafter. The design approach, of course, is not a magic formula. It cannot make good writing effortless. Nevertheless, the more writers understand about the writing process and the factors that influence their writing, the more easily they can make their writing decisions and the more effective their writing is likely to become.

Obviously, there is no one way to go about writing. Each person attacks writing problems in a way that differs somewhat from the approach of another. Yet a general writing process underlies these individual approaches, and certain basic principles must be observed. The design approach attempts to encompass these common elements.

CHAPTER 2

The Design Approach to Technical Writing

In advance of a detailed look at each stage in the writing process, this chapter presents an overview of the entire process and the design approach to it. Such an overview serves several useful purposes. Certainly it should indicate the basis on which the remainder of the book is organized. Then, too, a general understanding of the process as a whole can help in seeing the significance of its parts.

Even more important, perhaps, is the possibility that simply discovering the overall writing process can be helpful in itself. Many engineers and scientists find writing a frustrating experience because they are unsure about where to start and what to do next. In these circumstances it helps to recognize that, underlying the many ways in which writers go about their work, a series of tasks is more or less common to all informative writing. In fact, it is the process not its result that carries over from one writing job to another. The report produced today, for example, will differ from that produced next week, but the process for writing each report remains largely the same. Thus improved reports come from increased understanding of the writing process and more effective use of it.

Finally, a view of the whole process without too much detail should make clear that it can incorporate design or problem-solving procedures. Interestingly enough, many scientists and engineers successfully apply the problem-solving procedures to their technical work but never apply them to their communication problems. In their technical work they analytically determine the factors involved and methodically proceed to completing their design or solving their problem. For example, engineers begin the design of a gear box for a particular machine by considering the "input" and the "output." They would clearly determine the factors involved in the problem—torque, speed, and power ratios—then work out their solution. However, it may never occur to them that they can go about solving their writing problems in much the same way. Actually, in both cases the same principles and procedures should be applied. (See Figure 3, p. 7.)

THE WRITING PROCESS

In engineering design the quality of the product depends on:

1. Accurate and extensive analysis.
2. Complete and thorough investigation of alternatives.
3. Detailed and purposeful design.
4. Careful and ordered application of the design.

These basic steps are equally applicable to writing problems. In writing, as in engineering, each is important. The particular items to be considered in solving writing problems need further explanation, however, for they can represent somewhat unfamiliar material to a great many technically trained people.

Analyzing the Communication Problem

What is the purpose of the report?
How will it be used?

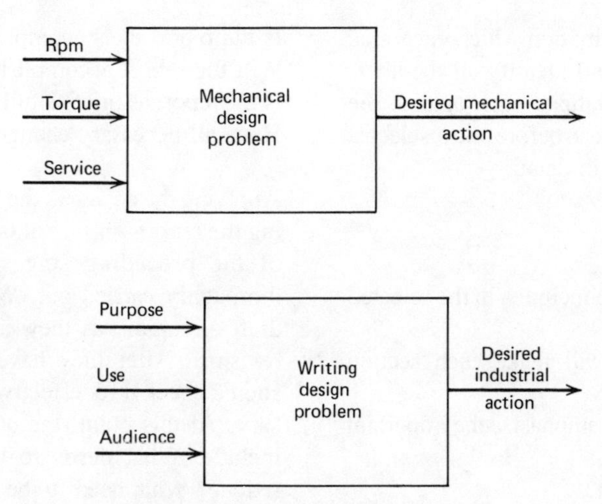

Figure 3. Similar design problems.

Who will read it?
What are the reader's informational needs?

The point of attack in report writing, as in engineering, is an analysis of the problem. Usually, analysis follows a two-step process: first, formulating the problem in broad terms, then carefully filling in the details. In writing, the broad formulation consists of identifying the purpose, the use, and the audience of the report to be written. Each of these factors greatly influences the kind of report to be written—the content that must be included, the organization of that material, and the style of the report. Before writers can prepare an effective and functional report, they must identify these factors accurately to determine how each will influence or limit the writing they are going to do.

The detailed analysis develops the implications of purpose, use, and audience that should affect preparation of a report. Writers must consider, for example, how the purpose they want to achieve and the informational needs of an audience are likely to determine the content of a report. They must recognize how their knowledge relates to the decision-making process of a given audience and how this relationship should affect content and its organization. They also must recognize how a report's probable use will affect the manner in which they might "package" their material and how the background of an audience will limit the language they use. These and many other implications of purpose, use, and audience require attention if writers are to understand fully what kind of communication problem they face and what they must do to work out an effective solution.

Investigating the Alternatives

What criteria must the report meet?
What major areas of content must be included?
What are the alternative patterns of organization?
Which pattern best meets the criteria?

Once the analysis is completed and the influence of purpose, use, and audience clearly identified, writers have the task of investigating the various organizational patterns available to them. With the analysis as a guide, they formulate the criteria they expect a report to meet; they identify the main areas of content they will have to include and gather the information so that it is at hand; they search for the alternative patterns of organization at their disposal; and they predict how well each may meet the established criteria so that they can select the best alternative.

Writers need to develop the habit of looking for alternative ways of structuring their reports, letters, or memos. There is no one way to do so. The task is to find the "best" way—the most functional way of presenting the material. Too many writers are content to use the first organizational pattern that suggests itself to them. Most frequently this pattern is a chronological or a work-structured one, and all too often it tends to obscure the information of greatest interest and usefulness to the reader. The choice of organizational pattern should be a conscious one, based on a careful appraisal of the possible alternatives.

In identifying the alternatives, writers should concentrate on the major areas of content and not get lost in considering the detailed information which each area will include. Also, they should identify all the alternatives open to them and evaluate each against the criteria established for a report before they select a pattern to use or work it out in detail.

Designing the Report

What detailed content must be included in the selected pattern of organization?
What internal organization will make each section most useful?
What internal layout will best emphasize the important material?
What illustrations are needed?

After considering the alternative ways of organizing their material and choosing the way best suited to solve a specific problem, writers are then ready to design a report. At this stage of the writing process, they should develop a detailed outline, expanding each major area of the overall pattern of organization they have chosen. In doing so, they also will be selecting in much greater detail than before the content the report must include.

The completed design should reflect, of course, the results of analyzing the communication problem. An effectively designed report is organized to fulfill its purpose. It presents material essential to the informational needs of the readers and supports this material with sufficient but not overburdening detail. The organization must ensure clear communication to those who are to read the report, must meet the requirements of the industrial use expected of it, and must provide the greatest usefulness to the reader.

All parts of the report must be made accessible to the reader by functional and effective use of headings. The layout of the report must reveal and support the organization of the content by presenting a graphic extension of the internal organization. Emphasis must be given to the most important material by devoting more space to the significant ideas, by placing them at the beginning of the report, and by effective use of location, spacing, headings, illustrations, and underscoring. The well-designed report achieves its purpose through functionally organized and properly emphasized essentials.

Applying the Design

Are the content and organization functional and effective?
Is the report clear, complete, and accurate?
Will the report accomplish its purpose?
Is the report as brief as full understanding will allow?
Have all necessary changes been made?

Applying the design in the fourth stage—actually writing the report—need not be difficult if the preparation of the preceding stages has been carefully and thoroughly carried out. Writers should create a first draft as rapidly as they can without worrying about revising. After they have drafted their report they should check it for effectiveness and revise it as necessary. At this point, the questions writers should ask include: What needs to be cut? What needs to be redone? What needs to be added? Is the report logical and clear, concise and direct? Are all sections accurately labeled and easily discernible? Will the report accomplish its purpose?

Abstracts and summaries should be written after the body of the report is finished, for listing the main ideas that now are actually in the report provides writers with the essential information for their abstract or summary.

Once the needed changes are made, a neat, accurate final copy is prepared. Copy is then reproduced and distributed to readers.

IMPORTANT FEATURES OF THE WRITING PROCESS

The view of the writing process presented here insists that, within reason, scientific and technical writing can be an orderly activity. It emphasizes that writing is not a single act, but rather a series of acts or tasks; it isolates the separate tasks involved; and it arranges them in a sequence of stages in which each task can best be accomplished. For the writers who have difficulty in getting started, it points out where they should begin; for the writers who seem to get "off the track" while writing, it can redirect their efforts.

Although the process as described sets out each stage neatly and cleanly, there obviously is some overlap from stage to stage in actual practice. For example, analyzing the communication problem not only is the starting point of the process, but also is a continuing activity throughout the process. Writers must return to the analysis of their problem, refining it as best they can when they investigate alternatives, when they design a report, and when they check and revise the first draft.

Often, a process when described in some detail can

appear to be long and complicated, but an important characteristic of this one is its flexibility. It is a process that can be easily telescoped for short pieces of writing. For example, before writing or dictating a letter, writers may spend a few moments thinking about what they want the letter to accomplish, who is going to receive it, and what the reader needs to be told. They may even jot down on a piece of paper two or three words that represent the topics they want to cover, and they may assign such topics a specific order before they start to write or dictate. Writers who prepare for composing a letter in this way are following the design approach to technical writing, although they may not take more than five or ten minutes in the writing of the letter.

The approach also insists that good writing requires time and that if writers are serious about their writing, they need to be realistic in scheduling their time for the various tasks that need to be accomplished. Deadlines, of course, should be realistic, but, even more important, writers must begin early enough on the writing process so that they will have enough time for its several tasks. No one has discovered a way to take the work out of writing or to reduce the amount of time required to do it.

Finally, this approach implies that each writing problem is a unique one and should be treated as such. However similar they may be, no two communication problems are ever exactly alike. Consequently, no stereotyped formula can solve each of these problems satisfactorily; there is no single format, no one organizational pattern, no homage to ''little words'' that can be used universally. Some kind of design effort in the writing process is basic to effective scientific and technical writing, whether the message be short or long, standard or nonstandard, and whether it be written for the scientist, the manager, or the general public.

The remainder of this book develops the design approach more fully. Each of four succeeding chapters takes up in detail a separate stage in the process. Throughout, the discussion emphasizes (1) factors that should be considered in designing a report to meet the requirements of a specific situation and (2) the tasks to be completed in writing a report.

CHAPTER 3

Analyzing the Communication Problem

Although scientists or engineers may have excellent solutions for their technical problems, unless they solve their writing problems effectively their work is of little value either to themselves or to others. Solving writing problems is generally difficult for all people. Writing is never easy, and technical writing poses many difficulties of its own. Certainly, an analysis of each writing problem is necessary. Rarely does the best solution arise by accident or chance. Before they can produce any report expected of them, writers must know why it is to be prepared and what it is to accomplish. If the report is to be effective, they also must be aware of how it is to be used. It is each writer's responsibility to present a report that can successfully meet the demands of the way in which it will be used. Finally, if the report is to be fully understood, it must be written so that its readers not only will comprehend its content but also will realize its significance. Although writers might wish it otherwise, they must assume responsibility for the reader's understanding of what they have written.

The analysis of any report-writing problem centers on the purpose, the use, and the readers (or audience) of the report, and it is on these that the technical writer should first concentrate in an effort to formulate the problem. Next, the writer should analyze the implications of these factors for the report to be prepared. This detailed analysis of the problem determines and controls what the writer must do—what content to include, how to organize this material, and how to present it. In reality, the analysis determines what report the writer should prepare.

PURPOSE

Reports differ as their purposes differ. Their scope and content are largely the result of the decisions that depend on them; thus each report must be designed to accomplish its own particular purpose.

How Reports Are Affected by Purpose

Writers should realize that the purposes of their reports must be their first major consideration. They should recognize, for example, that a report which reviews the actual observance of safety rules in a particular department will differ from one that aims at improvement and extension of the rules themselves.

The following three reports, growing out of the same industrial situation, differ because each was written to meet its own specific purpose. The first was written to report information, the second to initiate action, and the third to make recommendations.

Providing Information

In Report 1, Mr. Sears, the plant safety engineer, presents the findings of the electrical safety inspections to the supervisor of the Towel Production Department, Mr. Downs.

This report on the observance of the electrical safety regulations briefly and concisely presents the results of the inspections. This was its purpose, for whatever action is to be taken is the responsibility of the supervisor of the Towel Production Department. On the

Report 1. Observance of Regulations

Western Paper Company

Interoffice Memorandum **Date:** April 7, 1974
To: R. F. Downs
Supervisor, Towel Production Department
From: R. D. Sears
Plant Safety Engineer
Subject: Observance of Electrical Safety Regulations, Towel Production Department.

The following first-quarter safety report, presenting the findings of periodic inspections made between January 1 and March 31, 1974, is submitted for your information. These inspections reveal that, in general, your departmental personnel are observing the electrical safety regulations governing the adjustment, repair, and maintenance of equipment. Three instances were discovered, however, of failure to tag switches properly when two or more crews were working on the same equipment.

The following table summarizes the safety inspections made during the last 3 months.

Summary of Safety Inspections

Safety Regulation	Insp. 1	Insp. 2	Insp. 3
1. General Equipment			
a. 25–10 hp max. 400 volts opening and tagging switches	OK	OK	OK
b. Over 100 hp max. 400 volts opening and tagging by electrician only	OK	OK	OK
2. Paper Machines			
a. Switch opened by electrician only	OK	OK	OK
b. Notification of supervisor	OK	OK	OK
3. Tagging of Equipment			
a. Tagging by single crew	OK	OK	OK
b. Tagging by multiple crew	NO	NO	NO

R. D. Sears

cc: L. J. Ross, Plant Supervisor

Report 2. Stricter Observance of Regulations

Western Paper Company

Interoffice Memorandum **Date:** April 12, 1974
To: R. F. Downs
Supervisor, Towel Production Department
From: L. J. Ross
Plant Supervisor
Subject: Better Observance of Electrical Safety Regulations, Towel Production Department

In reviewing the last four quarterly Electrical Safety Regulations Inspection Reports for your department, I have become disturbed by the failure of your personnel to properly tag equipment when it is being worked on by more than one crew. The reports indicate that on only three of the twelve inspections made during the last year has the multiple tagging regulation been observed.

I fully realize the pressure of time exerted on the crews during a breakdown. Although time is important to us in our 24-hour-a-day production schedule, we must find some way of making the men realize that their safety is far more important to us than the time they can save by rushing when repairing an equipment failure.

At times safety is difficult to sell. You might find it worthwhile to discuss the problem of selling this particular rule with your group leaders and with Ray Sears. Quite possibly Ray could tell you what some of the other departments have found effective. We have not had a safety orientation program for the last two years; perhaps it is now time to start another one.

I know that you are aware of the danger to crews if they do not all properly tag the equipment when working on it and that you will take steps to see that the safety of your personnel is protected.

L. J. Ross

cc: R. D. Sears, Plant Safety Engineer

basis of what is reported to him, he must decide what is to be done.

Initiating Action

Mr. Ross, the plant supervisor, disturbed by these failures to tag the equipment properly, reviewed the

Report 3. Improvement of Regulations

Western Paper Company

Interoffice Memorandum **Date:** April 19, 1974
To: L. J. Ross
Plant Supervisor
From: R. D. Sears
Plant Safety Engineer
Subject: Recommended Changes and Additions to Electrical Safety Regulations

During recent discussions with the supervisor and group leaders of the Towel Production Department, three excellent suggestions were made concerning the plant electrical safety regulations. I recommend that the following suggestions be discussed at the next supervisors' meeting:

1. That rule 3, now reading, "Where more than one crew is involved on a job, it is the responsibility of each man equally to see that the equipment is properly tagged," be changed to read, "Where more than one crew is involved on a job, it is the responsibility of the foreman of each crew to see that the equipment is properly tagged."
2. That the following rule be added to the code: "Whenever the shutdown extends from one shift to another, it is the responsibility of the foreman of the crew (or crews) involved to report the status of the work to the foreman of the crew (or crews) on the next shift."
3. That the handles of all switches that are to be operated by plant electricians only (this would include the switches on all paper machines and all equipment over 100 hp) be painted a bright yellow, and that the following rule be added to the code: "All yellow switches are to be operated by plant electricians only."

These suggestions would considerably strengthen our safety code. The first would place the responsibility for proper tagging on one man rather than on the crew as a whole. In the past, equipment has not been properly tagged because each member of the crew left it up to the others. Although it was the responsibility of all members, it was in practice the responsibility of none. The second suggestion would greatly aid coordination between shifts and would ensure that faulty equipment is not turned on before it is completely repaired. The last suggestion would provide a clear and easy method of marking the switches to be operated by electricians only, eliminating the confusion that has at times existed in the past.

I join Mr. Downs and his group leaders in strongly encouraging the incorporation of these changes and additions into the present electrical safety code.

R. D. Sears

cc: R. F. Downs, Supervisor, Towel Production Department

reports covering the past year. Because the failures had been noted several times in these reports, he decided that he should exert some pressure to ensure fuller compliance with the safety regulations. On the basis of the last four quarterly reports, he wrote a memorandum to Mr. Downs (Report 2).

This report is confined to only one of the safety regulations, the one needing particular attention. In addition to pointing out that the rule is not being observed, the memorandum suggests steps to be taken in attempting to improve the situation. Written to accomplish its own specific purpose, this memorandum differs from the first report as its purpose differs.

Making Recommendations

After receiving the plant supervisor's memorandum, Mr. Downs held the suggested conferences with his group leaders and with Mr. Sears. In addition to deciding what action would be taken to correct the situation, the group leaders and Mr. Downs felt that three amendments to the safety regulations were needed. Agreeing that the changes should be made, Mr. Sears offered to write a memorandum to Mr. Ross recommending the changes (Report 3). Again, this report differs from the others as its purpose differs.

Summary: Plant safety rules are the subject of all these reports, but each one is different. The first includes a statement of past performance and records the extent to which the rules were followed or disregarded.

Report 1

Summary of Adherence to Regulations

1. Safety regulations followed.
2. Safety regulations disregarded.

The second stipulates the regulation that needs to be enforced and suggests steps to be taken.

Report 2

Stricter Observance of Regulations

1. The regulation that needs to be observed.
2. Suggested steps for finding ways to enforce the regulation.

The essence of the third report is a statement of modifications and additions to the present code, needed to ensure future safety.

Report 3

Recommended Changes and Additions

1. Modifications of existing regulations.
2. Addition of new regulations.

In each report the content and organization are largely determined by purpose. When faced with the task of writing a report, then, the writer must decide what the report is to do.

Classification of Reports by Purpose

Although many types of reports are used in government and industry, they can be classified according to their purposes, according to their relation to the decision-making process. Generally, reports are written:

1. To inform.
2. To initiate action.
3. To coordinate projects.
4. To recommend.
5. To provide a record of activity.

Of course, the relation of a report to the decision-making process is often a complex one, involving a combination of the purposes listed above. For example, all reports must inform the reader of something, and none of the other purposes can be achieved without doing so. In each case, however, the emphasis—the way in which information is provided—will vary. If a report's primary purpose is to provide the results of a test, the data must be emphasized. If the primary purpose of the report is to submit recommendations based on the results of the test, the recommendations should be emphasized, and the data should have a secondary role, that of supporting the recommendations.

In many reports various sections are designed to serve one of these purposes, but the overall organization and treatment should be governed by the primary purpose of each report. One section of a report might record the test activity: equipment, procedure, data, and results. Another section might set forth the recommendations that grew out of the test. The overall organization and development should, however, emphasize the recommendations because they are most vital to the decision-making process.

An understanding of the various purposes of reports will aid the writer in determining the kind of report to write or the kind of sections to include. Knowing the relationship of a report to the decision-making process helps the writer to know what to include and how to organize it.

To Inform

Those reports primarily designed to present information vary greatly in form, content, and length. By and large, they are of two kinds: (1) periodic reports presenting routine information and (2) special reports presenting information on a particular project, problem, test, experiment, or investigation. In both kinds of reports, however, the emphasis is on the information to be included. If the report is to be effective, the writer must clearly understand what information is needed and why.

In the periodic report the writer seldom has to determine its basic content and organization, because these are usually predetermined. Such reports are called for periodically, and the information to be included is routine. Thus up-to-date information is "plugged" into a routine organization. Often these reports are printed forms, and the writer fills in the blanks.

On the other hand, each special-information report requires a particular content and organization of its own. The information needed for a decision on whether to buy a new piece of equipment differs, obviously, from that needed for a decision on whether to install a new manufacturing process. Thus no standard prescription of content and organization can serve the writer from report to report. The problems of determining what is needed in each case, selecting the necessary information, and presenting it in the most usable way are basic to this kind of report; the success of each report depends on how well the writer solves these problems.

To Initiate Action

Reports written to initiate action usually do not leave much decision making to the reader; normally, they proceed from decisions already made. Their content focuses on answering the basic question of who is to do what in bringing about a change or beginning a new course of action. Almost always it also is important to make clear why the action is necessary or desirable and to state, or at least to suggest, how the action is to be carried out. Frequently, the reader may need information about the when and the where of the action.

In any case, the report that will get the action the writer wants must clearly designate the specific responsibilities of the persons involved, and it should include such details of why, how, where, and when that will make those responsibilities understandable. If the action called for is a complex one, the report should clearly present a detailed step-by-step procedure, and it should attempt to foresee and to suggest solutions for problems that are likely to occur.

To Coordinate Projects

In action reports the precise timing of events may not be overly important. However, in reports intended to coordinate the activities of several persons or groups of persons, the schedule showing when each group is to perform its job can become the critical item of information. If the trucks carrying concrete to a new highway site arrive before the roadbed is complete, the faulty or misunderstood schedule can be very costly

Often, coordination reports require that writers summarize the progress on a project and then estimate the times at which other facilities and personnel must be ready to take up succeeding phases of the project. If such reports are to have any value, their content must be up-to-the-moment information that bears directly on the activities to be coordinated. Time estimates must be as realistic and accurate as the writer can make them. Above all, other items of content should not be permitted to obscure the clear identification of who will do what and *when*.

To Recommend

Recommendation reports state what is recommended and why. Contrary to the belief of many engineers and scientists, recommendations must be sold, and an element of salesmanship is a part of any effective recommendation report. Selling recommendations depends on the writer's effectively presenting the need for their acceptance and the advantages to be gained. The *why* of a recommendation report is the selling part. The advantages of a particular recommendation may be obvious to the writer, but the chances are that they will not be obvious to readers. The latter will want to know what they should do, why they should do it, what they will get from it, and how much it will cost for what they will get. They will want to know whether the effects of a recommendation are certain or merely possible. They will want to know what risks they will run if they accept or reject recommendations.

Although recommendation reports vary widely, to be effective they must (1) present the recommendations themselves and (2) sell them by indicating their value to the organization, by telling the readers why these recommendations should be accepted.

Many recommendation reports describe the entire activity from which the recommendation grew. In such cases, the writer must take care that the presentation of the material does not smother the recommendations and that they do not appear tagged on at the end of the report. The recommendations and their value must always be foremost to be effective.

To Provide A Record

Reports written to provide a record of industrial activity are usually long and comprehensive; to be useful, records must be complete and detailed, thorough and accurate. Two types of reports are written to provide records: progress reports and final reports. For example, at the end of each year of work on the Oakland–San Francisco Bay Bridge, a progress report was written to present the scope and detail of the year's work. At the completion of the project, a report was written, covering the work accomplished since the preceding year's report. The several volumes of progress reports, each covering a particular time period, combined to make up the total record of the project. The final report on the San Francisco Golden Gate Bridge, however, is a one-volume report prepared after the job was completed. Drawing on other reports and records written during the construction of the bridge, it presents the entire project in retrospect.

Both methods are advantageous. The first tends to preserve detail and completeness; the second often provides a perspective impossible before the completion of the work. Perhaps the best system is to preserve facts and details in journal form during the project and then to rework them into an integrated whole after the work is finished.

Because readers often use these records long after the work has been done, the report must be comprehensive. The writer cannot safely assume that this reader is familiar with the project. Too much time may have elapsed; too much information may have faded from memory. Recording reports are part of an organization's history and must be written for people who were not involved in the project.

The purpose of a report exerts a strong influence on *the choice of a report's content, the organization, and the items that require emphasis*. Whatever the type of report, it has its particular purpose, and writers must clearly recognize what this purpose is. They must decide which kinds of content are major items to be stressed and which are subordinate items to be placed in the role of supporting detail. They must know what data to include and how to interpret their data for the reader. If conclusions are called for, they must know how acceptance of such conclusions would benefit the reader or the company. In any case, they must "sell" their ideas, not by promotion but by being fully aware of their ideas' value and making this value apparent to others. Their reports must communicate effectively, and doing so depends to a large extent on how well writers understand the purpose of a report and its relationship to decision-making processes.

USE

In addition to its purpose, the particular way in which a report will be used also should affect what it finally will be like. Consequently, the use to which a report will be put and the environment in which it will get used represent an important and necessary influence on the writing process. In analyzing a writing problem, the scientist or engineer needs to think about what will happen to a report from the time it is typed or printed until the time it is shredded, burned, or otherwise disposed of. More specifically, the writer needs to consider such questions as what kinds of readers will use the report, how many of them are there, how will they treat or handle the report, under what conditions, and for how long.

Reports may be used in many ways. Some, for example, may go to a single person inside one's own organization, be read only once, and then be filed. Others may go outside the organization to a rather large group of persons who will read these reports several times and even make them the focal point of decision-making conferences. Still other reports, serving as records or references, may have any number of readers who will consult them repeatedly for specific bits of information. Indeed, such reports may require a durable binding so that they can withstand considerable handling and long storage on library shelves.

The concept of use, then, emphasizes the variety of ways in which readers approach and use written material. Whatever the piece of writing, it should be designed to meet the use requirements of its readers. The use to be made of a report can exert influence on (1) organization; (2) devices for identifying content, such as headings and subheadings and the use of referencing systems; (3) choice of form; and (4) the physical characteristics of a report, such as reproduction method, use of color, type of paper stock, and sort of bindings. Perhaps the best way to explore the importance of use to the writer is to examine specific examples or problems.

Organization

Although the purpose of written communication and the nature of its audience usually exert the most influence on its organization, use can sometimes dictate how writers should structure their subject matter. The most common example of this fact is, of course, the familiar dictionary, which inevitably is organized in an alphabetical order. Theoretically, dictionary makers could organize what they have to say about words by putting all the nouns in one section, all the verbs in another, and so on. Quite obviously, however, this organization would not serve a dictionary's users, or readers, nearly so well as the alphabetical one. Because a dictionary is put together alphabetically, its readers can locate quickly the particular word about which they want information, obtain that information, and then put the dictionary back on the shelf until the next time they need it.

The same consideration of how use should affect organization applies equally well to reference manuals. Although these are not, strictly speaking, reports, scientists and engineers are frequently called on to write them. In doing so, the writers should consider very carefully how their readers are likely to use these manuals. On occasion an alphabetical arrangement of content may serve best. On other occasions such an arrangement may not be practicable. In these cases writers need to devise an organization of content that will become quickly apparent to readers and will permit them to locate specific items of information with a minimum of time and effort.

So far as reports are concerned, there is at least one type in which use can seriously affect organization. Quite often engineers and scientists must report on the status or progress of their projects so that their individual reports may be used in a much larger report representing them all. In effect, the individual reports are used to serve a broader purpose and set of audience requirements than the writers themselves might ordinarily concern themselves with. This fact may very well influence even the content of the individual reports, but certainly how and where the individual reports are to be used in the larger report are important factors in deciding how they should be organized. If writers are not made aware of or do not consider these factors, then the usefulness of their individual contributions is not likely to be very great. Indeed, the great difficulties that the editors of the inclusive reports often have arise simply because there is no general agreement on or understanding of how the individual contributions are to be used.

Identifying Content

In general the headings and subheadings which are an almost universal feature of technical reports serve as an aid to reading and understanding what the report has to say. Headings and subheadings are generally included to reveal the segments or sections of a report's content. Heading layout is a particularly important guide for readers, because it assists them in using the material a report contains. It should be emphasized, however, that the layout of headings is an extension of the organization of the material which is given a graphic representation on the pages of a report. The structure and position of the headings and subheadings call the attention of readers to the specific segments of information that make up the content of the report and indicate the relationships of those segments. They help readers to use a report by making material easily identifiable and easily accessible and by directing attention to those parts of the report the writer considers most important.

Frequently, writers find it necessary to assist the reader in identifying specific parts of a report or to provide a means for referring the reader to specific items in the report. For example, a construction design report may require that the builder be able to locate the dimensions of a support arm and its mounting holes. The presentation of such material in one section identified with an effective heading or subheading can make the information easy to locate and to use. The reader's use of a report usually requires that the writer provide specific labels for different items of content. In a report providing manufacturing design information and tolerances, the writer may find it useful to provide a paragraph or section reference system so that the manufacturer can be referred to specific requirements or specifications in later correspondence or reports. Frequently, a decimal reference system is used for each paragraph, each group of specifications, or each section and subsection, for example, 1.0, 1.1, 1.2, 1.3, 2.0, 2.1, 2.2. Such a system allows the writer to refer the reader to specific items of information in the report clearly and with precision.

Choice of Form

The choice of form is also related to use, because correspondence and report forms are essentially vehicles for carrying messages. Each is designed to handle a specific load and to establish a specific relationship between writer and reader. In effect, each report or correspondence form is a container, a container for information and ideas. Furthermore, the basis for choosing a particular form is much like that for choosing any other kind of container. Will it handle the amount and complexity of the material it will have to carry? The correspondence forms, memos and letters, and their short-report counterparts are small containers designed to carry a small amount of material. The article report form, on the other hand, is a medium-sized container, and the formal report form, of course, is a large one. It has the most organizational devices—Table of Contents, for example—and hence the ability to handle large amounts of material.

At the same time, the correspondence forms are relatively direct and informal means of sending messages. They permit the establishing of a close relationship between writer and reader; in fact, letters and memos are normally addressed to a specific reader or small group of readers. In contrast, the article report and the formal report are addressed to larger and more general audiences. Thus each form establishes a particular kind of relationship between the writer and the reader or readers.

Physical Characteristics

The method of reproduction is usually an economic decision based on the distribution of a report, the number of copies required, and the desired reader impact. Printing is expensive, and consequently offset processes are more often used when many copies are needed. If a report is to be used internally in an organization and is to have a limited distribution, it is frequently typed on bond paper, and copies are provided by using a photo-copy duplicator such as Xerox or 3M copier. Annual reports of companies, however, are widely distributed to stockholders and potential investors. Because of the desired reader impact, such reports are not only typeset but done in color on high-quality paper—technical reports seldom receive such treatment.

The experience of a West Coast aerospace firm illustrates how failure to consider use can create difficulties. The firm chose a plastic spiral binding for its manual describing the operation of a particular piece of equipment being delivered to the Air Force. The plastic spiral binding does have advantages. It permits a book to lie flat and pages do not flip—a decided advantage in a cookbook, where a disaster results if pages flip while the cook is in the middle of a recipe. However, one can safely assume one requirement for operations manuals for military equipment: once the equipment and manual are delivered, the equipment will go through a series of modifications, and the manual will have to be brought up to date. Thus the use of an operations manual for military equipment requires that someone periodically take out pages and put in new ones. Unfortunately, the plastic spiral binding does not readily lend itself to the insertion of new material, and, in light of this requirement, the choice of such a binding was a poor one. Had the writers seriously thought about the requirements of the environment in which the manual would have to be used, they would have recognized the need for updating it and consequently would have chosen a binding that minimized the problems of adding new material.

The use of a report, then, exerts influence on writers and on their writing, and, if they are to provide the best product they can, they must take these influences into consideration. Whatever the specific influence in a given writing situation, writers should make every effort to foresee the use requirements on their writing and prepare a product that can serve effectively in the environment in which it must be useful.

AUDIENCE

Just as purpose and use influence the nature of the report to be written, so does its audience—its reader or readers. To put the matter simply, writers must write for their readers. If they are to write effective reports, they must determine as best they can the reader's informational needs and background. The informational needs of the reader control, to a large extent, what a report should include and what should get emphasis. The reader's background in the subject of the report determines the manner in which the writer may present information.

Informational Needs

Engineers and scientists write for a variety of audiences, of course, but two groups of readers are particularly important—technical management and professional colleagues. Although much too little is known about either the informational needs or the reading habits of these readers, effective communication actually requires either a good knowledge of both groups or some very accurate assumptions concerning them. Fortunately, two studies conducted by J. W. Souther shed some light on the informational needs and reading habits of these two groups of readers.

Technical Management

Scientific and technical reports do not always convey information efficiently to management because the writers do not understand precisely what management wants to know. As a result, management readers have to winnow out the information they need or go back to the author with a list of specific question—and either procedure wastes time.

Recognizing this problem for what it was—insoluble without information on the kind of reports management wants—Mr. H. C. McDaniel of Westinghouse organized a study so that his company could get this information through personal interviews with persons in supervisory and management positions.

The agreement among the persons interviewed about what they wanted in reports was surprisingly high. Thus the results of the study provided a much clearer definition of management's informational requirements than had generally been available heretofore. In many instances, what previously had been assumptions about management readers were confirmed by solid evidence. This study was able (1) to identify the informational needs of management with considerable accuracy and (2) to determine how management uses reports and what its reading habits are.

The study identified five broad technological areas of primary interest to technical management:

1. Technical problems.
2. New projects and products.
3. Experiments and tests.
4. Materials and processes.
5. Field troubles.

Of course, managers want to know a number of things about each of these technological areas. Moreover, as the lists in Table 1 (below) reveal, the information that they want varies with each of the areas. In addition to the five technological areas, the manager also must frequently consider organizational problems and market factors. Although such problems and factors may not be a primary concern to engineers and scientists, they should include the information in reports going to their management. The types of concern that management has in these areas are included at the close of Table 1.

Although the questions in Table 1 are self-explanatory, two aspects deserve comment. First, the lists of questions pinpoint more accurately than heretofore the actual informational needs of the management reader. Second, the questions themselves point directly to the necessity of relating the technical report to industrial decision making. This is an important concept, one too often overlooked in advice on technical writing.

This emphasis on the report and its relation to de-

Table 1. What Managers Want to Know

1. Technical Problems

   What is the problem?
   How did it arise?
   What is the magnitude and importance of it?
   What is being done about it? By whom?
   Why was this undertaken?
   What approaches are being used?
   Are these thorough and complete?
   What are the suggested solutions? Which one is best? Should others be considered?
   What needs to be done now? Who does it?
   What are the time factors?

2. New Projects and Products

   What is the potential?
   What are the risks?
   What is the scope of application?
   What are the commercial implications?
   How stiff is the competition?
   How important is this to the organization?
   How much work must be done? Are there any problems?
   Are additional manpower, facilities, or equipment required?
   What is its relative importance to other projects or products?
   What is the life of the project or product line?
   How will this affect the organization's technical position?
   Are any priorities required?
   What is the proposed schedule?
   What is the target date?

3. Experiments and Tests

   What was tested or investigated?
   Why was this done?
   How was it done?
   What did the test show?
   Are there better ways to conduct tests?
   What conclusions do you draw from the test?
   What do you recommend be done?
   What are the implications of these tests to the company?
   What do you propose be done now?

cision making may well force some engineers and scientists into a realization they would prefer to avoid. The results of the study clearly indicate that presenting facts, the detailed data, is not sufficient. Data must be interpreted. Their implications must be revealed. In short, judgment becomes the focal point of the technical report. True, the judgment must be based on objective study and evaluation of the evidence, but it is judgment nevertheless. In fact, managers seldom use the detail, though they often want it available. It is the professional judgment of the scientist or engineer that managers want to tap.

Although the kind of information managers need is determined by their managerial responsibilities, how they want this information presented is determined largely by their reading habits. The Westinghouse study revealed that the report reading habits of managers are surprisingly similar.

When managers receive a report, they first want to know whether they should "read it, route it, or skip

4. Materials and Processes

    What are the properties, characteristics, capabilities?
    What are the limitations?
    What are the use requirements?
    What are the area and scope of application?
    What are the cost factors? Major?
    How available is the material or process?
    Is there another material or process that will accomplish the same result?
    How and where is the material or process best used?
    What are the problems in using the material or process?
    What is the significance of the application to the company?

5. Field Troubles

    What equipment was involved?
    What was the trouble that developed?
    How much money and time are involved?
    Is there any trouble history on this product?
    Whose responsibility to repair and place in service? Organization? Others?
    What action is needed?
    Are there any special requirements to be met?
    Who does what?
    What are the time factors?
    What is the most practical solution?
    What action do you recommend?
    What product design changes do you suggest?

6. Organizational Considerations

    Is it the type of work the organization should do?
    What changes will be required? Organization? Manpower? Facilities? Equipment?
    Is it an expanding or contracting program?
    What suffers if we concentrate on this?

7. Market Considerations

    What are the chances for success?
    What are the possible rewards? Monetary? Technological?
    What are the possible risks? Monetary? Technological?
    Can we be competitive? Price? Delivery?
    Is there a market? Must one be created?
    When will the product be available?

it.'' They need to know immediately whether the report has any bearing on the decisions they have to make. To determine if it does, they need answers to some or all of the following questions:

What's the report about and who wrote it?
What does it contribute?
What are the conclusions and recommendations?
What are their importance and significance?
What's the implication to the company?
What actions are suggested? Short range? Long range? Why? By whom? When? How?

Managers want this information at the beginning of the report and all in one piece, usually in a summary. Given this information, they can quickly decide whether they should read the report, route it, or skip it.

This is to say that, if a summary is to be effective, it must restate the essential information of the report. The problem, if there is one, must first of all be defined, and the objectives of the project set forth. Next, the reasons for doing the work must be given. The conclusion and the recommendations should, of course, be stated. The report summary must also contain a definitive statement about the significance of the work, followed by an interpretive statement of its implications. If it is appropriate to do so, the report's summary should tell who is to do what, when, and how.

Such summaries are informative and useful and will meet the informational needs of most management readers if they come at the beginning of reports. If a report does not include a separate summary, the same information should be presented as the first part of the report's main body.

All managers interviewed in the Westinghouse study said that they read the summary or abstract; almost all said that they read the introduction and the background sections, as well as those sections containing the conclusions and recommendations, in order ''. . . . to gain a better perspective of the material being reported and to find an answer to the all-important question: What do we do next?'' The remainder of the report was rarely read by those interviewed, and then only because a manager was

1. Especially interested in a particular subject.
2. Deeply involved in a project.
3. Forced to read by the urgency of a problem.
4. Skeptical of the conclusions drawn.

Those few managers who read the appendix material did so to evaluate further the work being reported.

To the report writer, this information on how managers read can mean but one thing: if a report is to convey useful information effectively to the management audience, the report's structure must not ignore the reading habits of this audience.

Technical Staff

The results of a second study—an effort to determine the informational needs and report-reading habits of engineers and scientists at a research-and-development laboratory of a national industrial organization—were a bit more surprising than those of the first. When asked what they looked for in a scientific or engineering report, members of the technical staff ranked their informational needs in order of importance as follows:

Items Most Often Looked For	Weighted Scale
Conclusions and recommendations	79
Statement of the problem	76
Approach used	62
General concepts	58
Special problems	50
Results	45
(and at the bottom of the list)	
Detailed data	16

This ranking is surprising, because in talking with scientists and engineers at the time they are writing their reports, one may get the feeling that they consider the results and the data as the important part of the work that they have to report. When questioned about what their readers want, they usually conclude by saying, ''Well, at least the scientist or engineer (the only reader who counts) is interested in the data and the results of my study.'' Yet, when scientists and engineers were asked what they looked for as readers, they attached little importance to what they usually consider as most important when they write.

This contradiction, when examined in greater depth, reveals two basically different kinds of reporting (and reading) situations. One is a tandem reporting situation in which the work is handed from one group to another as the project moves, for example, from R & D to Preliminary Design, to Design, to Manufacturing, and so forth. In tandem reporting the detail of the report is extremely important, because the next group must have it in order to carry on the project. But this need for detail is more the exception than the rule in reading reports. More often, the technical reader is working on problem x and finds a report by another scientist or engineer who was working on problem y. What the reader wants is some assistance in solving problem x, and by looking at the report on problem y, the reader hopes to find something that will help with his or her own problem.

Now what is there about one problem that might be useful in solving another? Conclusions, approach, general concepts?—of course. The interest in the statement of a problem reflects an effort to relate another person's problem to the reader's problem to see how much alike they really are. Special problems may be of interest because they can call the reader's attention to some things that must be looked into but have not yet been foreseen. Results and detailed data, however, are not very transferable, for they grow out of the specific problem being reported on and are not likely to fit any other problem.

Let there be no misunderstanding. Detailed data cannot be left out of a report. They need to be included, but the data usually should not be the focal point of the report. Rather, they should support and substantiate the important concepts and conclusions that have grown out of the work. Data should be presented in the perspective of the work being reported on.

Although they only scratch the surface, these two studies provide a partial guide to audience analysis by identifying the informational needs and reading habits of two different groups for whom the engineer or scientist so often writes. The results of both show that writers should emphasize the larger and more important aspects of their work and put their detailed data in a supporting role.

Writing for the Reader

Although it is often difficult to put oneself in the place of a reader, this technique is most valuable to the writer of technical reports. When the reader is a person whom the writer knows, the problem is least difficult. When many persons will read a report, the problem of audience analysis becomes more complex, even if the writer knows each individual. When the readers include persons whom the writer does not know, then analysis of the audience must rest on rather critical assumptions, and the report will be no more effective than are the writer's assumptions concerning the audience of the report. Whatever the case, whether the audience be one person or several, whether the audience be known or not, report writers must either find or assume answers to three main questions: (1) What does the audience want? (2) What else does it need to know in order to understand? (3) What is the audience's background? Whatever writers decide, their answers to these three questions will affect all that they do and all that they write.

What Is Wanted?

Giving readers what they want—what, in many cases, they have asked for—is fundamental to industrial and government writing, and those reports that do not do so fail. When department heads request information or professional judgment from members of their staff, they want it to aid in making decisions. If they do not get the information or advice they want, they cannot make a decision, or more likely and even worse, they make it without sufficient evidence.

Providing readers with what they want is the primary obligation of report writers. This they must do if they are to succeed. Often, however, technical writers are so concerned with what they want to tell the reader that they never consider what the reader wants of them and the reports they are writing. These reports succeed only if both the writer and reader "see eye to eye," and such an approach to technical writing is extremely dangerous.

For example, if a decision must be made about which of two new conveying systems should be installed, the factors of greatest importance are service, dependability, and cost. If in comparing the two systems an engineer becomes particularly interested in the many new features of one of the systems and emphasizes these in a report, management will not get the information needed for making a decision. Although the new features may be advantageous, they may also be sufficient reason for rejecting the equipment. It is possible that these new features might require new skills of the maintenance crews and that additional people might have to be hired to provide the skills. Would this involve more union contracts? Would there be difficulty in determining which work should be done by personnel of one union and which by members of another? If the system is new, would parts be locally stocked? Or would time have to be wasted in getting them from a distant parts house? Although the uniqueness of the design may well have attracted the engineer, the manager still wants to know: What kind of a job will it do? How dependable is it? How much will it cost? The system that does 10 percent more work at 50 percent more expense might well be rejected for a system that provides less service at a more reasonable cost.

In any case, the writer should first make certain to provide what the reader wants to know. A report may include additional information, but it should never do so at the expense of what the reader wants.

What Else is Needed?

In addition to providing what readers want, report writers must also include all the additional information that readers will need to understand fully the answers they expect to get. To achieve this end is no easy task, and writers will be successful only if they can foresee what content will be necessary for a full understanding of the reports they are to write. Thus, drawing on their own experience and their analysis of each writing situation, they must do their best to identify the specific kinds of material they should include.

By raising and carefully answering the question of what a reader needs to know, writers can help themselves in two ways. First of all, they can more specifically define the content of their reports than they could if they were to stop with the question of what the reader wants. Essentially, the answer to "What does the reader want?" indicates the major subject areas of a report, whereas the answer to "What does the reader need to be told to fully understand what he or she wants?" indicates subdivisions of these major areas.

In the example of the two conveying systems, the answer to the first question would suggest that a report should cover three general areas:

1. Comparison of the service provided by the various systems.
2. Comparison of the dependability of the various systems.
3. Comparison of the costs of the various systems.

The answer to the second question, however, suggests a set of subareas for each of the general ones:

1. Comparison of the service provided by the various systems.
   a. Capacity.
   b. Safety.
   c. Flexibility.
2. Comparison of the dependability of the various systems.
   a. Data from tests.
   b. Data from the operational experience at similar plants.
3. Comparison of the costs of the various systems.
   a. Initial costs.
   b. Operating costs.

Perhaps writers can, and should, define their subject matter in even greater detail during the analysis stage. Of course, the more they do so, the more likely will be their need to modify their original definition as they go on to gather and evaluate the actual material. On the other hand, the more detailed their definition, the less likely they will be to overlook what is essential to a reader's understanding.

Asking what the reader needs to know can be of help in yet a second way. It can suggest necessary areas of content that might not be so obviously related to what the reader wants. Very often the technical writer, meeting a specific assignment, tends to equate the reader's wants with what the reader has asked for. This, of course, is not entirely a bad principle; at least it might prevent writers from including too much irrelevant material in reports. On the other hand, it can also cause them to overlook something essential. The writer of the conveyor-systems report, for example, when asked for information about two systems, might very well have stopped with the topics listed above. Yet by carefully following through on the question of what the reader needs to know, this same writer might have found these topics inadequate. What good, the writer might legitimately ask, will a comparison of service and dependability be without the requirements that the conveyor system must meet? How will the reader know if one or both systems meet the requirements or exceed them to an unnecessary degree? And how can the reader choose one of the two systems without this information? Thus, thinking of the reader's needs, the writer might add another general area to the original list:

1. Comparison of the service provided by the various systems.
2. Comparison of the dependability of the various systems.
3. Comparison of the costs of the various systems.
4. Requirements that the various systems must meet.

Some technical writers might object that managers making such a decision would know what requirements the conveyor systems should meet and that to tell them what these requirements are would be unnecessary or presumptuous. Yet it is a common observation of those busy readers, the managers, that they seldom recall at a moment's notice all the information they are supposed to have. Consequently, on many occasions reminding managers of what they should know can be serving their reading needs. Certainly writers must exercise judgment in this matter. However, writers also should make certain to provide not only what readers want but also whatever else they need to know in order to understand fully the material and its significance *at the time they are reading*.

The Reader's Background

When writers know who will read their reports, they should consider how the backgrounds of their readers may limit what they write. If they do not know their readers, they must assume a background and write accordingly—usually to persons of intelligence and good general knowledge, but without much experience in the specific subject of a particular report.

Too often, engineers and scientists ignore the backgrounds of their readers. Instead, they write as if no one should read a report except another engineer or scientists of the same breed, interest, and experience as the writer. No wonder many reports fail in communicating much to their readers. To produce successful reports, the writer frequently must work at translating technical ideas into nontechnical language. More often than not, little is lost and a great deal is gained.

Writers must determine the technical and mathematical knowledge of their readers. They must decide just how "technical" they can be. They must decide on the extent to which they can assume a reader's familiarity with a subject, for the reader's background and experience may limit the concepts the writer can use, the kind of illustration that will be clear, the detail that will have meaning, and the special terms that will communicate. It is always well to keep in mind that presenting ideas so that the readers will understand them is primarily the writer's responsibility. Writers cannot assume that readers will look up unfamiliar words or take steps to provide themselves with the education necessary to understand highly technical details. It is much easier to *not read the report*.

The study at Westinghouse also provided some information on the technical level of presentation that managers would like. Usually, these readers have backgrounds of education and experience that differ from those of the scientists and engineers writing the reports the managers receive. Moreover, as noted previously, these readers are essentially interested in the significant ideas that grow out of detail. Consequently, they are seldom pleased by a highly technical and detailed presentation. In any event, the technical managers at Westinghouse indicated that reports written for them should be at a level that would be suitable for a reader whose background of education and experience is in a field different from that of the writer. For example, the report writer who is an electrical engineer should write reports for a person educated and trained in chemical engineering, mechanical engineering, or metallurgical engineering—not in electrical engineering.

No general rules, of course, will ensure that writers have met their responsibilities in the communication process, but they will have gone far toward mutual understanding when they realize that the responsibility for the clearness of ideas is their own—not the reader's. They will write, then, to persons, if they know their readers. If they do not know an audience, they will take intelligence for granted and neither write over the reader's head nor treat the reader like a child. Aware of the importance of a reader's background, writers would not prepare the same report describing a proposed design for the chief accountant that they would for the staff engineer. In writing reports to be read by both the management and engineering branches of a company, they would provide the management personnel with a general summary of significant material in nontechnical terms and would include a detailed and more technical discussion for the engineers. Such a report serves both groups and serves them well, for it provides readers with what they want, supported by what they need to know, expressed in terms they can understand.

Summary

Realistically and clearly understanding the purpose, use, and audience of a report is the writer's first task. All these factors exert a variety of influences on the writer and writing. At times it is difficult to distinguish among the factors, and more than one of them may influence a particular portion or characteristic of a piece of writing. Purpose and audience frequently seem to work hand in hand, for both exert heavy influences on the selection of content, organization, presentation, and emphasis. Use, on the other hand, most frequently affects the physical characteristics of a piece of writing, including the choice of form, layout devices (such as thumb tabs), reproduction processes, binding, and paper. Yet each of the factors can influence all of these areas, and it is important that writers consider fully how each will affect their reports.

The initial analysis, of course, is not the last. So much depends on it that it must be returned to at various stages of the report-writing process. It must be reexamined, extended, corrected, and made more specific. The stages of writing are not entirely clear-cut; they fuse one into the other; and writers constantly find themselves reviewing their earlier decisions. Nevertheless, the initial analysis of the writing problem is the starting point. The more thoroughly and accurately it is done, the more sure and effective will be the writer's work in the following stages, for it becomes the basic guide by which the writer works in preparing a report.

EXERCISES AND DISCUSSION TOPICS

1. Identify the purpose of a report or paper you have written and describe how your selection, organization, and presentation of its content related to and supported that purpose.
2. Examine two or three engineering reports from industry or government agencies like ERDA or NASA. Chose one to discuss, and answer the following questions:
   - **b.** What is the purpose of the report? How does its organization reflect its purpose?
   - **b.** What was the intended use of the report? What evidence did you find to indicate that it was designed for use?
   - **c.** What would you assume to be the nature of the audience for whom it was written? How does the report reflect the writer's awareness of his or her audience?
   - **d.** What changes would you suggest?
3. Select a report, an article, or a book, and discuss how its intended use has influenced its organization, layout, and form.
4. Examine several articles in two or three journals or magazines directed at (1) engineers or scientists, (2) management, and (3) the general public (e.g., *Chemical Engineering, IEEE* publications, *Science, Scientific American, Popular Mechanics, National Geographic, Business Week*). Choose two articles that represent attempts to communicate with two different kinds of readers, and compare the organization and levels of presentation in the articles. Comment on how each article meets the informational needs and interests of its readers, and include any suggestions that you may have for increasing the effectiveness of the articles.
5. Choose a subject with which you are familiar—for example, the operation of an automobile clutch, the function of a carburetor, prestressed concrete, a chemical process, the lift produced by a wing, the operation of a TV picture tube—and write two explanations: (1) to a friend who is a freshman engineering student and (2) to another friend who has no technical training or background. (If you wish to attempt a difficult task, write the explanation to a 10 year old.)

CHAPTER 4

Investigating the Alternatives

After analyzing the writing problem, the writer is ready to investigate the alternative ways in which to present the kinds of material a report should contain. Investigating these alternatives requires that the writer develop criteria for the report to be written, identify the content that is essential, search for alternative patterns of presentation, and finally select the one pattern that will best serve the purpose and readers of the report. If the implications of purpose, use, and audience have been worked out realistically, then consciously investigating the alternative solutions for a given writing problem helps to eliminate unneeded work in which writers can so often get bogged down. Approaching the preparation of a report analytically focuses attention on the objectives to be achieved and then concentrates on what the writer must do in order to reach them.

DEVELOPING CRITERIA

The criteria to be used in selecting the overall pattern of organization for a report usually begin to surface during the analysis of the problem. The general criteria for writing, of course, actually change very little from one situation to another. All technical writing should be functional, readable, clear, direct, and as simple as possible in its presentation of ideas. However, the choice of alternative ways to present material does not rest so much on such general criteria as it does on the specific ones that grow out of each writing situation. Every report has its own blend of specific purpose, particular use, and unique informational demands to fulfill for its audience. Thus the writer should push the analysis of a writing problem to the point at which its distinctive features become clear and, from these, develop a list of criteria for the report to come.

Because useful criteria are so closely related to the specific writing situations out of which they grow, any discussion of their development and application has need of an example. Consequently, it will be helpful to return to the conveyor system project discussed in Chapter 3. In this example of a writing problem, an assistant engineer was asked to report to the plant engineer about two different systems—the gravity roller and the belt conveyor systems. Some rather simple analysis indicated that the plant engineer needed enough appropriate information about both systems and the requirements they must meet so that he could decide which system to recommend for installation in the plant. The purpose of the assistant engineer's report, of course, was essentially to provide this information. However, by pushing the analysis a bit further, the assistant engineer might have come to consider the nature of the decision-making process the plant engineer must follow in order to arrive at a recommendation. Presumably, the process is one of comparison—of comparing both systems with the requirements that must be met if the needed services are to be provided. At this point, then, the assistant engineer might well have concluded that at least one of the criteria for the report should be something like the following statement: The report's structure should permit a quick and easy comparison of both systems and the requirements they must meet.

Later, after identifying the report's content, the assistant engineer might have undertaken to evaluate the two following patterns of presentation of information about service:

Pattern A

1. Requirements
    - Capacity
        - Load
        - Speed
    - Safety
    - Flexibility
2. Roller System
    - Capacity
        - Load
        - Speed
    - Safety
    - Flexibility
3. Belt System
    - Capacity
        - Load
        - Speed
    - Safety
    - Flexibility

Pattern B

1. Capacity
    - Load
        - Requirements
        - Roller System
        - Belt System
    - Speed
        - Requirements
        - Roller System
        - Belt System
2. Safety
    - Requirements
    - Roller System
    - Belt System
3. Flexibility
    - Requirements
    - Roller System
    - Belt System

Faced with this choice, the assistant engineer now can make use of the criterion developed earlier. Evaluating the alternatives comes down to answering the question of which organizational pattern will provide the better basis for the comparisons the plant engineer must make in order to arrive at a recommendation.

Both patterns set up comparative relationships, but they do so in significantly different ways. Pattern A presents the information to be compared in large, separate blocks. Thus the reader must keep in mind all the requirements and the capabilities of the roller system while studying the capabilities of the belt system. Making comparisons in this manner is not a particularly easy task.

Pattern B, on the other hand, permits a point-by-point comparison of the two systems in relation to the requirements. The reader, for example, can focus on the load-capacity requirements, compare them immediately with the load capacities of both systems, and arrive at a judgment about the relative merits of each system as far as these requirements are concerned. Having done so, the reader can leave this point of comparison behind and focus on the next one, free of any need to remember much of what has gone before.

Of the two patterns, then, Pattern B promises to be more functional for the reader than Pattern A. It fits more closely the actual decision-making process the plant engineer will go through and, consequently, offers the easier way to arrive at a recommendation. Because it does satisfy the established criterion better than the other one, Pattern B should be the assistant engineer's choice for the report on the conveyor systems.

Unfortunately, there are no fixed rules for developing criteria, and those used for one report are not likely to be appropriate for the next. Criteria simply vary from one writing situation to another. Nevertheless, writers can and should develop the ability to create appropriate criteria for each piece of writing they do. In analyzing each writing problem, they can aim at setting specific goals to achieve in the finished piece of writing they are working toward. Implicitly, at least, the assistant engineer had as one goal for the conveyor-systems report, that it present the necessary information in a manner that would be most helpful to the plant engineer in reaching a decision about which of two systems was the better. Such goals, however, need translating into criteria on which to base decisions about what an effective report should be like. The imaginary assistant engineer came up with a useful criterion rather easily, but, with practice and experience, actual writers may do almost equally as well.

IDENTIFYING CONTENT

With specific criteria developed, the writer's next task is to identify as well as possible the content the report probably should include. Quite obviously, the writer should have the material to be shaped well in hand before seeking alternative patterns in which to present it. In this effort, it is wise to proceed in two steps. The first is to complete the definition of content that very likely began during the analysis of the writing problem. The second is to gather the information that this definition calls for.

Defining Content

As criteria can begin to emerge during analysis, so can the general outlines of a report's content. At least the latter normally will emerge if writers consider carefully the implications of their purposes and the informational requirements of their readers. How far the process of defining content goes during analysis varies with different writers and the particular writing problems they face. Usually, though, considerable refining of a content's definition remains to be done when a writer moves out of the analysis stage. Completing this unfinished business is important if the writer is to have an adequate (although certainly not infallible) guide for collecting the material that will be essential to a report. In each case the definition of content should be as full and detailed as the writer can make it in order to guard against overlooking some item of information that the reader ought to have.

Two primary approaches are available to writers in their efforts to define content. One, deductive in nature, proves workable whenever writers arrive rather early at a reasonably clear overall idea of what their reports should say. On these occasions, the major areas of content may come readily to mind, and the further definition of content proceeds by deducing what subdivisions of content each major area should logically include, what subdivisions each subdivision should include, and so on—always keeping in mind, of course, the report's purpose and audience.

Once again, the report on the conveyor systems is useful, serving to illustrate the deductive approach to defining content. In analyzing the problem posed by this report, the assistant engineer considered what the plant engineer wanted and rather quickly concluded that three major considerations in comparing the two systems would be service, dependability, and cost. Then, asking what else the plant engineer needed to be told, the assistant engineer deduced that the category of service, for example, should include information about capacity, safety, and flexibility.

Having gone this far, however, the assistant engineer had not yet defined in any detail the kinds of information required for a comparison of the service the two conveyor systems can provide. Some important questions remained to be answered: What kinds of capacity are involved? What kinds of safety? What kinds of flexibility? Thinking through the answers to these questions, the assistant engineer might have a broken down the original category of service in the following way:

- Capacity
  - Load
    - Total number of pounds
    - Total number of units
  - Speed
    - Range of speeds
    - Control of speed
- Safety
  - Of personnel
  - Of products
- Flexibility
  - For changes
    - Routine
    - Temporary
  - For addition of new facilities

At this point the process of identifying in ever-increasing detail what broad areas of content involve has probably reached its practical limit. To ask further questions about "range of speeds," for example, should lead directly to gathering specific data on the two conveyor systems. Of course, the assistant engineer needed to define equally well other areas of content, too, like those of dependability and cost. Yet the process would have been the same and the results similar. Certainly a critical look at "initial costs" might have suggested that they could include the purchase price of each system, its delivery cost, and a set of installation costs that would require further identification. Similar treatment of "operating costs" might also have suggested that, in addition to direct operating costs, there would be costs for maintenance and repair as well.

Perhaps the foregoing account of the deductive approach to defining content tends to make it seem a bit simpler than it usually is. Even so, pressed vigorously enough, the process of breaking broad areas of content down into ever smaller and more specific units works very well—provided one starts with some clear notions of what a reader wants and is likely to need. At times, unfortunately, such notions may not emerge so clearly from analysis of the writing problem as writers would like. Consequently, the general outline of the content needed in a piece of writing remains hazy, and writers have great difficulty in getting started at all in defining it.

On these occasions, writers should consider using a second approach, an inductive one. If they cannot proceed systematically from large units of content to smaller ones, then they can reverse the process. They can begin with such bits and pieces of content as they can identify and work toward constructing the general framework within which these fragments will fit.

The actual starting point, of course, is, again, the results of analysis. To write down the purpose of a particular report and the informational requirements of its audience is a good practice at any time, but doing so becomes especially helpful whenever writers must define a report's content inductively. The inductive process needs a focal point, and the best one available is a written statement about purpose and audience. Indeed, it pays the writer to spend some time in making the statement as full and specific as careful analysis permits. Then, with the nature of the purpose and audience firmly fixed in mind, the writer can proceed to each of these steps, in turn:

1. *List all of the ideas for content suggested by the report's purpose and audience requirements.* This is no time to be critical, to evaluate each idea as it comes. Rather, the point is to relax the critical faculties, to give the mind's association of ideas free play, and to put down quickly everything that one can think of, in whatever order and in whatever form.
2. *Review each item in the list critically.* Once the flow of ideas has definitely stopped, it is time to become critical. The task now is to clarify and evaluate the relationship of each idea to purpose and audience. The objective is to identify those ideas that are obviously relevant to a report's content, those that are so obviously irrelevant that they can be crossed off the list immediately, and those that lie somewhere between the extremes. Most frequently, the items in the last category are ideas whose relevance is somewhat in doubt; but caution recommends that a decision to eliminate them be put off until later in the process. Occasionally, an apparently irrelevant idea "insists" on remaining in the list and "forces" a revision of the original statement about the report's purpose and audience.
3. *Group related items.* The remaining items still represent nothing more than a jumbled list of ideas. However, a close inspection of this list will show that a few of the ideas it contains are more closely related in substance to one another than to any of the others; a second inspection will reveal a second such group; a third inspection, a third group; and so on. Thus the essential work at this point is to construct a new list in which all groups of closely related ideas are separated from one another. Particularly as the new list becomes complete, its developing pattern of groups may serve to guide decisions about the relevance of doubtful items held over from the second step. At the same time, obvious gaps in or between the groups may suggest new relevant ideas that did not get into the list of the first step.
4. *Construct a general framework within which the groups fit.* This activity begins with working out for each group of ideas a heading—a word or phrase—that indicates how the ideas in a group are related to one another. In addition, each heading should (and probably will, anyway) indicate how its group is related to the other groups, as well as to the report's purpose and the informational requirements of the audience. Almost always, however, the initial headings do not fully reveal all the significant relationships among the groups. The similarity of two or three headings, for example, usually means that their groups are more closely related to one another than to the other groups. Where such relationships show up, it is important to place the closely related groups under a broader, common heading:

   (Common) Heading
     (First) Heading
       Idea
       Idea
     (Second) Heading
       Idea
       Idea
       Idea

   Moreover, collecting the groups together under broader and broader common headings should continue until all the groups logically fit within a few broad areas of content. As this framework develops, additional ideas for content may very well come to mind, and remaining questions about the relevance of some items are likely to resolve themselves.

For a number of scientists and engineers, the inductive approach to defining content may seem much too disorderly a process to be very trustworthy. The advice at the first step, especially—to relax one's critical faculties while jotting down the original list of ideas—may appear as nothing more than an invitation to sloppy thinking. Nevertheless, this step has one great advantage: it does get potentially useful ideas down on paper, where the writer can work with them.

In listing ideas, the writer should not give up the first or even the second time that ideas stop coming. Association moves in "spurts," and the writer needs to give the process a chance to start up again. Moreover, the writer should resist the tendency to become critical while listing ideas, for such a shift in attitude can shut off the flow of ideas and severely reduce the usefulness of the preliminary list. Ample opportunity to become critical is provided in those steps following the listing of ideas.

Writers may find that using the inductive approach effectively may take a bit of practice. If they carry out each of its steps conscientiously, however, the result can be as useful a guide for gathering content as that provided by the deductive approach. Indeed, the results of both approaches should be quite similar: a small set of headings labeling broad areas of content that, in turn, are subdivided one or more times by less and less inclusive headings.

Moreover, if writers believe it helpful to do so, they can turn the result of either approach into a rough kind of outline. They can take this step simply by deciding on a possible order in which they might present to readers the broad areas of content and the various subdivisions of these. Indicating this order by means of appropriate arabic numerals produces something like this:

2 Heading (broad area)
  3 Subheading
    2 Subheading (Idea)
    1 Subheading (Idea)
  1 Subheading
  2 Subheading
    2 Subheading (Idea)
    1 Subheading (Idea)
    3 Subheading (Idea)
1 Heading (broad area)
  1 Subheading
  2 Subheading
    1 Subheading (Idea)
    3 Subheading (Idea)
    2 Subheading (Idea)

Of course, some writers may wish to go even farther and produce a new copy of this rough outline, rearranged in the order on which they have decided.

Normally, however, such an outline cannot serve as the final outline, or actual writing guide, for a report. The writer has yet to obtain the detailed data and information suggested by the headings; this material may call for additions to and possibly alterations in the rough outline. Also, the actual gathering of data is very likely to uncover unforeseen material that "insists" on being included whether there is a heading for it or not. Most importantly, a pattern of presentation set up at this point in the writing process will probably be the first pattern that occurs to the writer, not the one that will serve the reader best. As later sections in this chapter stress, the writer should consider some alternative patterns of presentation and choose the one that most effectively satisfies the criteria established in relation to a report's purpose and audience. It is premature to attempt this task before the writer has actually collected the data which a report requires.

Gathering Content

Gathering material, particularly in scientific and engineering projects, is usually complex and time consuming. On one hand, the plan for obtaining the necessary information should be carefully thought out, but, on the other, it should remain flexible to accommodate changes as the work of gathering proceeds.

If writers are fortunate, the information they need is already available. Their task, then, is to identify the sources and seek out the information they need for the report they are to write. In formulating an approach to gathering information already available, they need to make certain that all the potential sources are used. Writers should, of course, determine what they already know. Then, after clarifying their own thoughts, they can investigate the sources of information available to them including (1) other people working in the field, (2) departmental and organizational files, and (3) published material in journals, reports, and books.

They should, of course, know how to make a literature search. Engineers, scientists, specialists of any kind, should know the identity of the primary reference guides and indexes in their specific fields. They should also, of course, be familiar with library information coding systems, know how to locate reports on studies made within their own companies or agencies, and be familiar with the professional literature of their fields.

Most often, however, writers find that the information they need must grow out of the technical work they are assigned. At such times, the writer must solve both the technical problem and the communication problem described in Chapter 1. Both problems should be given attention from the beginning and throughout the work.

As the diagram in Figure 4 (see p. 30) shows, efforts to solve both problems can run concurrently, although the steps of each problem-solving process will not always be in phase.

The important point is that gathering of information for a report should go on as the technical work progresses. Too often, engineers or scientists concentrate on their technical problems to the exclusion of their communication problems. Consequently, they usually find themselves "writing by recall." It is no wonder, then, that writing is more frustrating than it need be. Alert

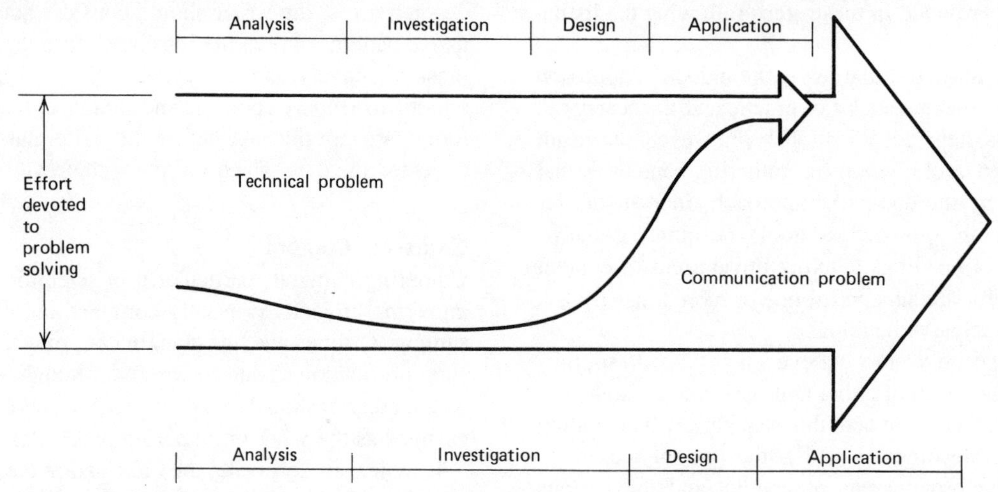

Figure 4. Interrelationships of problem-solving methods for projects involving both a technical problem and a communication problem.

writers collect information as they do the technical work; for example, when they arrive at a statement of the objectives of the technical work they are to do, they set it aside in a report file. They make a habit of looking out for items that they should set aside, such as a statement of the problem, a description of the approach used, and records of the major work done and results achieved and of the major ideas or conclusions that emerge during the work. Such a file provides reminders and information which they can evaluate and use later without having to depend solely on their power of recall once the work is completed.

Writers should also realize that gathering information for reporting must always be an active procedure, never a passive one. They should constantly keep alert for what fits their definition of the content they believe the reader requires. And, of course, writers are not seers; they can not always foresee every item of content that may be relevant to a report. However, if they have thought carefully about what is needed and are alert, they are likely to recognize the unforeseen items that a report, nevertheless, must include.

SEARCHING FOR ALTERNATIVES

Having gathered a body of material, writers should seek out alternative patterns for presenting it. That they shold do so comes as a complete surprise to many writers, for they tend to choose the first organizational pattern that occurs to them for structuring their content. This habit accounts for the overuse of work-structured presentations in technical writing, for writers are most likely to think of the order in which they solved their technical problems when seeking an organizational pattern for their reports. Yet if scientific and technical writing is to be effective, looking for other ways to present the material is essential.

There is no one way to organize any report. Alternatives are always available. The writer's task is to discover the best way, the most effective and most functional way, to structure a report, or, for that matter, a letter or memo. Determing the best way requires that the writer consider the available alternatives.

Some alternative ways of organizing content may occur to writers as they analyze their communication problems, but these alternatives are merely by-products of the analysis, not the result of a deliberate search. They should, of course, be identified at the time as possible alternatives. On the other hand, they simply offer a starting point of the conscious search for alternative patterns of presentation which writers should carry out. Finding the best pattern seldom results from happy accident; it usually comes from a thorough and thoughtful effort to identify all the possibilities for organizing a particular content.

At the same time, finding the best organizational pattern seldom occurs if writers let their search for alternatives be influenced too greatly by the traditional patterns that have grown up in their own company or agency. Because each new report differs from past reports, writers should use their own ingenuity and skill in problem-solving when they seek alternative solutions to writing problems. The extent to which they exercise their ingenuity depends, of course, on their attitude toward writing and the importance they attach to effective communication. If they simply wish to get the writing job out of the way, they are not likely to be very diligent or inventive in their search for alternatives. However, if they believe that communication is important and take report writing seriously, they will spend some time and care in seeking out alternatives and judging which of them might be the most effective pattern for a given report. They will assume not only that there is more than one way to organize a report, but also that they are likely to discover a better way than the one which first occurred to them, if they will but take the time to discover it.

A number of factors tend to reduce the effectiveness of writers in their search for alternatives. Too many writers depend on the traditional, commonly used patterns of presentation for each new report they write. Thus, at best, the organization of each report simply reproduces the strengths *and the weaknesses* of previous reports. Organizational prescriptions and established tradition are extremely strong influences, and writers who give in to this pressure without any thought of alternatives are seriously risking the quality and effectiveness of the writing they produce. Too frequently writers are more concerned about defending the standard report organizations they have come to depend on than they are about increasing the effectiveness of their communications.

Writers who want to seek out the alternative patterns for presentation of their writing, then, must be willing to expend the effort and exercise the inventiveness needed to identify the viable alternatives. In making this search, however, writers must avoid a series of traps.

1. *Avoid premature decisions.* Writers must make certain that they do not choose the first organizational pattern that comes to mind, but recognize that their task is to identify the various organizational patterns available to them.
2. *Avoid premature evaluation.* Seeking out possible alternatives usually stops when writers start to evaluate the alternatives prematurely. The search phase becomes mixed up with the decision phase, and once decisions start to be made, the search usually ends. Judgments should be made only after the various alternatives have been identified and their general outlines worked out.
3. *Avoid getting bogged down in detail.* If a writer works out the details of the first organizational pattern that comes to mind, for all practical purposes the search has stopped. Preoccupation with the details of one alternative usually interferes with the writer's ability to think about different patterns. The task here is to search out the overall patterns that might serve as alternative ways of presenting the material. They can be evaluated in a relatively undetailed state of development.

SELECTING THE PATTERN OF PRESENTATION

Once the alternatives have been identified, writers can then begin the process of selecting the particular pattern they believe best meets their criteria. Although this general decision-making process varies from one situation to another, three steps are usually involved: (1) predicting the performance of alternative solutions with respect to the criteria formerly determined, (2) evaluating the alternatives on the basis of these predicted performances, and (3) choosing the most effective. To determine which organizational patterns will best meet the established criteria, the writer must be able to predict to some extent the performance—the ability to communicate—of the various alternatives being examined. In any case, the task is to determine which of the presentations is most functional to the purpose and to the informational needs and decision making of the readers.

Perhaps the following example can help in understanding the kinds of considerations writers must take into account when they are evaluating alternative patterns of presentation. In this example an administrator of a sea-transport firm has just learned that one of his ships, the *Pandora,* has been involved in an accident at sea. His information is sketchy: The *Pandora* under temporary repairs is headed for Columbia Inlet, a small port where the ship does not usually call; it has suffered damage forward of amidships on the starboard side where plates were ruptured; water was taken aboard but after the temporary repairs the pumps are now handling the situation; no one was injured and there were no deaths; cause or responsibility for collision has not been determined; and the *Pandora* will arrive at Columbia Inlet in two days.

The administrator has moved quickly to make arrangements for taking care of the incoming ship. He has contacted the Phoenix Salvage Company, the only concern of its kind in the area, asking it to assist in unloading and salvaging cargo and to assess the damage to the ship. He has also determined that the

Columbia Inlet Shipbuilding Corporation is the only one available at the port and that Reliable Testing Company can be used to assess the salvageability of the cargo.

He is now faced with the task of writing instructions to his operations manager who is closest to Columbia Inlet, asking the manager to go there, take charge of the in-port operations, and report important information back to him.

In preparing to write his memo, he has worked out the following patterns of presentation.

Pattern 1

- Background information
  - Collision
    - Damage to plates
    - Water taken aboard—pumps can handle
    - No injury or loss of life
  - Arrival
    - Columbia Inlet
    - June 10th
  - You go there, take charge and make reports
- Cargo
  - Extent of damage unknown
  - Phoenix Salvage will assist in unloading and salvaging
  - Keep accurate records of cargo saved
  - Keep accurate account of water-damaged cargo
  - Keep record of cost of unloading and estimate difference over normal unloading
  - Send samples of questionable cargo to Reliable Testing Company
- Ship
  - Determine extent of damage
  - Phoenix Salvage will assess damage
  - Be sure they look at more than ruptured plates
  - Get bid for repairs from Columbia Inlet Shipbuilding Corporation

Pattern 2

- Background information
  - *Pandora* in collision at sea
  - No injury or loss of life
  - Heading for Columbia Inlet
  - Arrives June 10
  - You go there, take charge and make reports
- Damage
  - To ship
    - Location just forward of amidships, starboard side
    - Plates were ruptured
  - To cargo
    - Ship took water until temporary repairs were made
    - Water damage to cargo unknown
- Action
  - For cargo salvage
    - Phoenix will help unload and salvage cargo
    - Send samples of questionable cargo to Reliable Testing Company
    - Keep records
      - Of cargo saved
      - Of cost of salvage unloading
      - Of damage to cargo
    - Estimate difference between salvage unloading and normal unloading
  - For ship repair
    - Determine extent of damage
    - Phoenix will assess damage
    - Be sure they look at more than ruptured plates
    - Get bid on repairs from Columbia Inlet Shipbuilding Corporation
  - Report information to me as soon as possible

Pattern 3

- General information
  - *Pandora* in collision at sea
  - No injury or loss of life
  - Damage to ship and cargo
  - Arrival at Columbia Inlet June 10th
  - Go there and take charge of operations
- Action we have taken
  - Contacted Phoenix Salvage Company
  - Arranged for it
    - To assist in unloading and salvaging cargo
    - To assess damage to ship
- Action we want you to take
  - About Damaged Cargo
    - Unloading
    - Salvaging
  - About Damaged Ship
    - Thorough assessment of damage
    - Get bid from Columbia Inlet Shipbuilding Corporation
  - About reporting
    - Cargo saved and damaged
    - Cost of salvage unloading
    - Estimate of difference between salvage and normal unloading
    - Bid for ship repairs

In working out these three organizations for an instruction memo to his operations manager, the administrator included the same basic information in all of them. Moreover, all the alternatives are reasonably logical. The emphasis of each one, however, differs from that of the others. In evaluating them and selecting the one that would be most effective, he had to determine which would be most functional to his purpose and most useful for his operations manager. The first pattern emphasizes the interrelationships of the content items—of the subjects to be included. The second adds a stress on action to the information concerning damage. The third places emphasis on the supportive action the administrator has taken and the action the operations manager is to take. Because of its emphasis on ''people roles'' as well as on basic information, the administrator selected the third pattern for his memo. It does, in fact, allow for providing a kind of checklist for the operations manager's convenience.

Selecting the most functional alternative, then, depends on how well writers understand their reports' purposes and uses and the informational needs of their readers. Out of such understanding emerge the criteria for evaluating alternative patterns of presentation, for every piece of writing has its own special combination of specific purpose, particular use, and unique informational demands from its readers. With relevant criteria in mind, writers can select the most effective way of organizing the content they have gathered, if only they will consciously seek out and evaluate the alternatives available to them.

EXERCISES AND DISCUSSION TOPICS

1. Prepare a list of criteria for a report you are going to write for the class.
2. Use the inductive approach to define the content and potential organization for a report you are to write for class.
3. Develop two or three alternative overall patterns of presentation for the report you are going to write. Keep the emphasis on the arrangement of the main sections; don't work out your outlines in detail. Explain why you believe the one you intend to use is most functional to your report's purpose and to the informational needs of your readers. Explain how the pattern of presentation you choose meets the criteria you established for the report.
4. Examine a report or article and discuss the effectiveness of its overall pattern of presentation. Can you identify alternative patterns that the author might have used? Would any of the alternatives provide a better presentation? Explain.

CHAPTER 5

Designing the Report

Having chosen from the possible alternatives the overall organizational pattern that will be most functional, writers then face the task of working out a report's design in detail. At this stage of the writing process, writers must shift their attention from general consideration of alternative organizations to selecting all of the content they will actually include in a report and fitting it into the pattern they have chosen. The objective now is to prepare a full blueprint of the writing product. In carrying out this task, writers, like architects, should reconsider the specific demands that people and their needs will place on the finished report. Successfully detailing the design requires that writers keep foremost in mind the purpose, use, and audience that a report is to serve.

Although completely organizing a report is always an exacting and difficult task, it becomes much easier if writers have prepared themselves to do so. If they have followed the writing process by analyzing a problem, investigating alternatives, and choosing the most functional one, the shape of the report should have come into clear focus. Thus the task of filling out the design and formalizing it should have some rather sharply defined limits.

Performing this task requires three very different activities: (1) selecting and organizing the detailed content of a piece of writing, (2) providing graphic and tabular support through proper use of illustrations and tables, and (3) choosing the form, layout techniques, and physical characteristics for the finished product that are most appropriate and helpful. Frequently, inexperienced writers undertake only the first activity with any care and give less attention than they should to the second and third. Nevertheless, paying careful attention to all three is necessary to an effective written product.

SELECTING AND ORGANIZING CONTENT

The primary means of finally selecting and organizing content is to develop an outline that can serve as a detailed guide for writing the report. In effect, use of the inductive or deductive approach to defining content (described in Chapter 4) has already led to some rough outlining. Moreover, choice of an overall organizational pattern has established the broad areas of content the report will include and has also set up an order in which to present these areas to the reader. Now, however, the writer actually has at hand a body of data and information about which the following questions remain unanswered: Do each piece of data and each bit of information collected really belong in the report? If so, in which section of the report, in relationship to what other items of information, and in what order? By answering such questions, writers select and organize the detail required for the broad content areas they have already chosen. Providing the details that fill out the general topics is one of the writer's main responsibilities, for effective presentation of detail is a characteristic of good technical writing.

Beginning with the broad outline they have selected as the pattern for their presentation, writers may vary in how they select and organize the detail needed to complete it. Some may work deductively from general topics to the detailed subdivisions of each. Others may use the inductive method to detail the main sections of their outline. Still others may mix the two approaches, using the inductive one to identify the major subdivisions of a section and then working deductively to develop these major subdivisions further.

Constructing the Outline

In any case, writers find it important to refine their statement of the main (or overall) idea a report is to convey. Or, if the main idea has remained implicit in their thinking up to this point, they should state it in words now as clearly, concisely, and specifically as they can. This statement should crystallize the thinking they have already done about the purpose of a report and the informational requirements of the audience. It should reflect what they have learned in defining and gathering possible relevant content for the report. Obviously, the statement of the main idea also should embrace the broad areas of content that are included in the overall organizational pattern.

The writer of the *Pandora* memo, for example, might have phrased the main idea of the message to the operations officer in this way: "When the damaged *Pandora* reaches port, you are to follow up the action I have taken by determining unloading costs and the extent of cargo damage, by supervising the assessment of ship damage and getting a bid on repairs, and by reporting on these matters as soon as you can." On the other hand, the assistant engineer providing information about two conveyor systems to the plant engineer might have seen the main idea of the report as being: "A comparison shows that both the gravity roller and the belt systems will meet the plant's service requirements for capacity, safety, and flexibility; that both will be about equally dependable; and that the belt system will cost more to install, operate, and maintain than the gravity roller system." Later, incidentally, the plant engineer might have stated in the following way the main idea for a recommendation report to the plant manager: "I recommend that we install a gravity roller system in the plant, because although either the gravity roller or the belt system will meet our needs, the belt system will cost more to install, operate, and maintain than the gravity roller system."

Carefully and explicitly stated, such main ideas define the scope of each report's content. In doing so, they draw the broad areas of content into a coherent whole and may suggest ways in which to phrase more accurately than before the headings with which writers have labeled those areas. More importantly, the statement of the main idea sets limits on the kinds and, to a degree at least, the amounts of data and information that each broad area of content should contain. It represents a sketch of a report's boundaries. As such, it can guide the writer in reviewing and evaluating the material gathered for a report and in finally selecting the detailed content of that report.

The outline a writer constructs as a writing guide should be a logical development from the main idea and its major divisions, represented by the headings that label the main areas of content. How carefully the writer constructs such a guide depends on the circumstances. For some pieces of writing—usually short, simple, and routine—an informal or scratch outline is sufficient. Memos and letters, and very often short reports, normally require no more than a scratch outline. In these cases, for example, the product of the inductive approach to defining content described in Chapter 4 may serve well enough.

Although a scratch outline is adequate for many pieces of writing, there are times when writers should prepare a formal outline (topic or sentence). In doing so, they should avoid the temptation of simply dressing up a scratch outline with the convention of designating main divisions with Roman numerals, their subdivisions with capital letters, and so on. Instead, the formal outline should reflect the care needed to phrase topics clearly and accurately and to be sure that the logic of division and subdivision is not flawed by missing, overlapping, or unwarranted topics.

Frequently, writers have need of the formal outline when they must write some important correspondence or a significant report. Almost inevitably they need one whenever they face:

1. *Long and complex pieces of writing*. Large amounts of material and its inclusion, ordinarily, of an increased level of detail call for a formal outline.
2. *Delay in writing*. If there is to be a delay in writing, it is best to translate a scratch outline into a formal one in order to retain the full meaning of the various topics. The hastily phrased ideas of a scratch outline can lose meaning quickly.
3. *Submittal of an outline*. When writers are asked to submit outlines for review, they should provide formal outlines because these usually communicate ideas more effectively than the informal type. Prepared in a writer's own "shorthand" for expressing ideas, the scratch outline often means very little to anyone else.
4. *Cooperative writing projects*. If two or more persons are to write parts of the same report, the formal

outline can serve much more reliably than the rough scratch outline to define with precision what each writer is to cover.

Selecting Material

Whatever type writers use, the completed outline should include enough division, subdivision, and sub-subdivision of its major headings to specify rather fully what a report's content will be. Just how many levels of division an outline should have is a difficult question to answer in any general way. The number depends on the probable length of a report, the complexity of its subject matter, its purpose, the informational requirements of readers, and the care with which each person wishes to prepare for writing a first draft. Nevertheless, a frequently heard piece of advice is that each topic at the lowest levels of division in an outline should represent the subject matter of a paragraph. In practice, writers seldom achieve such a goal with great accuracy. At the same time, considered as a rough rule of thumb, the advice is worth heeding, for many writers stop far too short of its goal when constructing their outlines. Indeed, writers often resist the idea of developing an outline in sufficient detail. As a result, they put off too much selection of content until the time when they are drafting a report, and they add an unnecessary complication to an activity that presents difficulties enough. Thus it will repay writers to spend time and care in carrying the topical subdivision of an outline as far as they possibly can. The more closely they can approach the specific data and other information a report should include, the more clearly will they establish what its detailed content will be, and the fewer decisions must they make as they labor at writing the first draft.

It may mean very little, for example, to set down the topic "test data" and stop there, if the body of test data gathered for a report is rather large. Too many decisions about content may remain unrecorded in the outline. Will the report contain several types of test data, and if so, what are they? Must the report include every one of these types? Do some types of data have two or more significant aspects—the results from a variety of tests, for example? And is it necessary that the results of every test go into the report?

Provided these questions are the appropriate ones, their answers represent decisions that finally get down to making a selection from the specific material the writer has gathered. Moreover, the fact that such questions remain unanswered clearly indicates that selection of content has not gone far enough when the writer stops at the topic "test data." At this point, the outline should include additional subtopics—perhaps some like these:

1. Test data
   a. Type I
   b. Type II
      (1) Results of test a
      (2) Results of test c
   c. Type III
      (1) Results of test b
      (2) Results of test d

Unfortunately, choosing what is significant for a report from material one has gathered can also mean discarding other parts of the same body of material. Consequently, selecting the material to be used can be difficult for many writers. After all, they have gathered this information in the process of answering a question or solving a problem. Often the material has been difficult to obtain, and discarding hard-won information is not a pleasant task. Furthermore, writers may have gathered their data with an eye to serving the purpose of a report and meeting the informational needs of its readers. With some justice, then, they may well ask why now they should seriously consider abandoning material that has cost them at least something to gather.

Put most bluntly, the reason is that the act of gathering it does not guarantee that material will be relevant for a report. In seeking the answer to a question or the solution of a problem, scientists and engineers may collect data that finally prove irrelevant to their purpose. With few exceptions, such data are doubly irrelevant when the time comes to report their answer or solution to someone else.

If writers do not discriminate in their choice of content, if they do not determine what is significant, their reports will be difficult to read and to use. Important ideas become smothered by less important detail. Relationships become fuzzy, and the reader has difficulty in comprehending the material. Such writing, at best, becomes a rough collection of information from which the significant must still be mined. Its writers simply expect their readers to do the discriminating work that they should have done. Consequently, the reader sees the report as disorganized and difficult to understand. The clear communication of ideas depends, in a large part, on how well writers distinguish between the main ideas and the supportive detail and on how well they present these relationships in their reports.

In selecting material, writers face two basic decisions. First, they must decide what detail goes into the report and what is to be left out altogether. Once the decision to include certain information has been made, writers are then faced with the decision of where it should go—into the main body or into the appendix.

The choice between putting information in the body or in the appendix should not be a casual one. Rather, it should be based on the role the information is to play in the report, or its relationships to the report's purpose and the reader's informational needs. If the content is essential to understanding, it should be placed in the body of the report. If it supplements general understanding by presenting supportive detail that might be useful to a few readers who seek in-depth understanding, it should be placed in the appendix. Many reports would become clearer and easier to use if writers would make this distinction in the role of content. If the content is necessary to the reader's general understanding, it should go into the body; if it provides in-depth understanding or back-up detail, it should go into the appendix.

Organizing Material

In their efforts to organize content, writers are really preparing ideas for the reader's consumption, and they must keep the reader's needs foremost in mind while organizing. Frequently, a report can be organized effectively around a set of questions readers might ask. For example, if the purpose of a report is to get support for a specific recommendation or a course of action, the questions basic to a decision-making process might well form the organizational pattern. In writing such a report to managers or the general public, the writer might do well to organize the report around the following sequence of questions:

Why me?

> Let the readers know from the beginning why the report is of concern to them. Identify their involvement.

What's it about?

> Provide the readers with a view of the purpose, scope, and nature of the report. Identify what will be covered.

What should I do?

> Clearly state the recommendation, the decision, the course of action for the reader.

Why?

> Although this is the shortest question, it requires the longest answer. Here the writer presents the reasons, the evidence, the rewards and risks, the advantages and disadvantages of the recommendation or desired action.

What alternatives are there?

> Identify and discuss the other actions that could be taken, pointing out their advantages and disadvantages, including how they fall short of the choice recommended in the report.

This sequence of questions forms the framework for many successful recommendation reports because it anticipates the natural flow of the reader's questions. Like any organizational pattern, it will not fit all situations. If, for example, the subject matter of the report is highly controversial or if the report recommends a course of action toward which the reader is not receptive, the pattern needs to be modified because the flow of questions will be different. In such cases, the alternatives might better be presented before the recommendation.

Whatever the situation, writers should keep their purpose and their reader's informational needs clearly in focus as they organize their material. When they can, they should anticipate the questions that will most likely occur to the reader and provide the information necessary to answering them.

In organizing a report, the writer is really working out its actual content, for until the organization is carefully worked out, the detail that makes up its content is never clearly identified. But organizing is more than identifying content; it is also the establishing of relationships and order. These are the ingredients of effective organization.

Many writers view organization only in terms of overall patterns. Although the overall pattern is crucial to effective organization, other matters need attention. Organizing a report also consists of developing internal patterns of presentation for the various sections. Actually, several different patterns are frequently used within the overall pattern. Writers need to develop a sophisticated view of organization which includes both the overall pattern and its internal variations.

Writers of environmental impact statements, for example, usually follow a prescribed pattern of organization, one established by the law itself. Does this mean that such writers have no real organizational problems? Quite to the contrary, such reports present difficult organizational problems. The problems, however, have to do with presenting material effectively in the specified sections rather than with the overall pattern.

Establishing Relationships

Writers need to establish carefully in their outlines the patterns of relationships that exist among the facts and ideas they are going to present. Not only must they see these relationships clearly; they must express the relationships so that readers will also see the same ones. The need for making relationships clear can best be seen by examining the following example from a report concerned with a section of forest.

Specific Use

1. Livestock management—salting and fencing. Trucks are generally used to haul horses over system roads to a point near the salt area. Salt may be distributed in some areas with four-wheel-drive vehicles. Little or no damage is caused by this use but may have contributed to some of the existing nonsystem roads.
2. Hunting—Four-wheel-drive vehicles are likely to be seen in any area that is not closed because of slope, timber, or rocks. This travel is generally during the wet season, resulting in rutting and severe damage.
3. Sightseeing—Travel is generally over maintained roads. No damage results.
4. Vehicle "venture"—This type of travel by car is generally over existing roads that are well traveled. However, the increasing use of trail bikes for nonsystem travel is creating rutting and severe damage.
5. Post and Wood Cutters—This type of travel is generally over main and logging spur roads where material is easily reached. Little or no damage is caused by this use.
6. Fishing—only about 5 percent of the travel is over nonsystem roads. This travel is generally during dry summer months, so little damage results.
7. Prospecting and Mining—This type of use has resulted in most of our poorly constructed nonsystem roads. Following construction, these roads receive extensive use by other forest users. This excessive use and poor construction has resulted in extensive damage in the North Fork drainage.
8. Power line maintenance—This is accomplished by travel over system roads. No damage is created by vehicle travel.
9. Timber Operators—This use is generally over recently developed spur roads and main roads. Reconnaissance work is done by travel over any road that is safe for passage. This does not create any damage.
10. Recreation (mining cabins)—Travel is generally over old nonsystem roads during the hunting season. This often results in badly rutted conditions.

How are these items organized? What relationships among them did the author intend to convey? The layout of the items in a list, with emphasis on specific land uses, leads readers to believe that this was the organizational intent of the author, but was it? Another pattern suggests itself because of the frequency of comment concerning amount of erosion damage resulting from these uses. Which pattern of relationships was intended?

KIND OF USE

COMMERCIAL USES

- LIVESTOCK MANAGEMENT
- POST AND WOOD CUTTERS
- PROSPECTING AND MINING
- TIMBER OPERATORS
- POWER LINE MAINTENANCE

RECREATIONAL USES

- HUNTING
- FISHING
- VEHICLE "VENTURE"
- MINING CABINS
- SIGHTSEEING

AMOUNT OF DAMAGE

LITTLE OR NO DAMAGE

- LIVESTOCK MANAGEMENT
- POST AND WOOD CUTTING
- TIMBER OPERATORS
- FISHING
- SIGHTSEEING
- POWER LINE MAINTENANCE

SEVERE DAMAGE

HUNTING
PROSPECTING AND MINING
VEHICLE "VENTURE"
MINING CABINS

The readers, then, are faced with making a choice as to which they think was the set of meaningful relationships intended by the author. Unfortunately, the readers can only guess which of the two possibilities was intended. Had the author organized this list by providing significant groups of related items, the readers could immediately have grasped the intended pattern. This example, much more typical than many writers would admit, reveals what happens when writers fail to organize their material carefully. The report's author probably put this section of the report together from rough notes in a list, and the unstructured listing carried through into the writing. For the readers' sake, however, it would have been much better to group the items so that their significant relationships were readily apparent. What may be clear to a writer will not be clear to readers unless the writer makes it so.

Ideas and facts gain significance from their interrelationships, and writers need to express these interrelationships if readers are to understand them. Being aware of them is not enough. They are not obvious unless stated, and writers often fail to include these important relationships unless time is taken to group related ideas.

Establishing Order

In addition to working out significant relationships in their outlines, writers also need to be concerned with the variety of orders available to them in communicating their ideas effectively. Some of the more commonly used orders include the following:

NATURAL ORDERS

SPACE—material organized from a spatial point of view. Used in describing physical objects.

TIME—material arranged chronologically. Used in describing a process, an event, a procedure.

LOGICAL ORDERS

CAUSE TO EFFECT—material organized on a causal relationship, from cause to effect or from effect to cause.

GENERAL TO SPECIFIC—material organized on a basis of generalization and support. Can be from general to specific (conclusion to evidence) or from specific to general (from evidence to conclusion).

IMPORTANCE—material organized on its significance. Can be in order of ascending importance (climax), or descending importance (anticlimax), or a combination of these—from high point to detail and on to climax (brokenback).

COMPLEXITY—material organized from simple to complex.

FUNCTION—material organized on the basis of its application. Moving from axioms to proofs or from criteria to comparison of alternatives.

FAMILIARITY—material organized in order to take the reader from familiar to unfamiliar knowledge.

ACCEPTABILITY—material organized in order to take the reader from an acceptable position to one that might not otherwise be acceptable.

These are only a few of the organizational patterns available to writers, but they are all frequently used in technical writing. All are useful; yet some are overused and some are too seldom used. The order of time is probably the most frequently used order. It certainly is the most widely misused order in technical writing. When engineers and scientists who have just completed an assignment are asked to write it up, they are most likely to start at the beginning and structure their writing on the order in which they did the work. Yet this, in many instances, is an ineffective organization, particularly for a management audience. Although the order of time is an excellent one for telling someone how to do something (describing a procedure), it is a poor structure for reporting the results of one's own work.

If chronology is overused, the general-to-specific order tends to be underused in technical writing. Yet this order is one of the most useful for informative writing. Readers can move easily from a generalization to the supportive detail, because this sequence of ideas keeps material in perspective for them as they read. Provided with a generalized view, they are prepared for what follows. There is, consequently, less chance that they will become confused or lost in reading a report.

As these virtues of the general-to-specific order may suggest, however, an important point in constructing an outline is to choose an order for presenting content that makes it most accessible and understandable to the reader. None of the orders listed above nor any other can serve best as the overall sequential pattern for all reports. At the same time, one or more sections of a report very probably will require an order of ideas different from the one the writer has selected for the report as a whole. Thus in working out their detailed outlines, writers should strive not only to indicate in each part all of the significant relationships of their material but also to arrange the various subdivisions in an order than can guide the reader to the understanding necessary for achieving a report's purpose. Clarifying relationships and selecting functional orders takes time for careful thought, of course. Nevertheless, the time can be well spent, for the result is most likely to be a clear and effective guide for writing a clear and effective report.

Revealing Organization

The fact that the outline is a writing guide leads to yet another aspect of efforts to organize content. It is important for writers to have a full and clear view of their pattern of presentation; yet much of their effort to reach this view will be wasted if, in writing a report, they do not explicitly make their pattern clear to their readers. As Bruce Sanford of the U.S. Fish and Wildlife Service has pointed out, one of the major problems in effective communication arises from the failure of writers to recognize how much their view of their material as a total unit (with sequences and logical relationships of coordination and subordination) differs from the straight-line sequential view of their readers. Their multidimensional view, as expressed in their outlines, has both breadth and depth, sequence and relationships, as shown in Figure 5. (See p. 41.)

What is the readers' view of the same material as they read? Is it as full and coherent as that of the writer? Far from it! The reader's view, by the very limitations of the reading process, is one that differs greatly from that of the writer.

Readers get their information line by line, sentence by sentence. Information comes to them one idea at a time, one behind the other. They receive ideas in *sequence,* but the writer's view contains more than sequence. It has coordination and subordination as well. These intricate relationships are often the very key to the meaning and the significance of the material for readers.

Although this difference in view presents a real problem in much technical writing, its solution lies in writers' recognizing that two kinds of content are necessary to clear, understandable reporting: *informative content* and *organizational content.* Informative content consists of the ideas and information in the report, the content that is organized for presentation during outlining. Although such informational content is always included in reports, too often the organizational content is omitted. In the following example, the writer proceeded immediately from a section heading (''Parts of a Letter'') to the first subdivision of the section, in which he describes the heading of a letter. He knew where he was going, but what about the reader?

PARTS OF A LETTER

Heading

The heading is usually a printed letterhead in business correspondence, at times however, it is typed. In either case the heading furnishes the reader with a name and address of a company or the address of the writer; the date when the letter is written is added by the typist. When correspondence is filed the envelopes are thrown away and the reader has only the letter as a record of information. Therefore, the address of the writer or of his firm is essential data that must be included in the letter itself. The heading of a letter then serves this function by providing the name and address of the company or of the writer and the date that the letter was written.

Actually, this subsection is followed by others in which the writer describes the inside address, the salutation, the message, the conclusion, and the complimentary close. The reader, however, is not aware of where the writer is headed, or how the material is to be developed and why. How much clearer it would have been for the reader had the writer inserted the following short paragraph between the first heading and the subheading.

All business letters contain a heading, an inside address, a salutation, a message, a conclusion, and a complimentary close. Each part has its own particular function.

This paragraph tells the reader (1) what the subject of each subdivision of the section is, (2) what the order of presenting the subdivisions is, and (3) what each subdivision contains—a description of the part and a statement of its function.

This map paragraph, a piece of organizational content, leads readers into the section. It orients them to

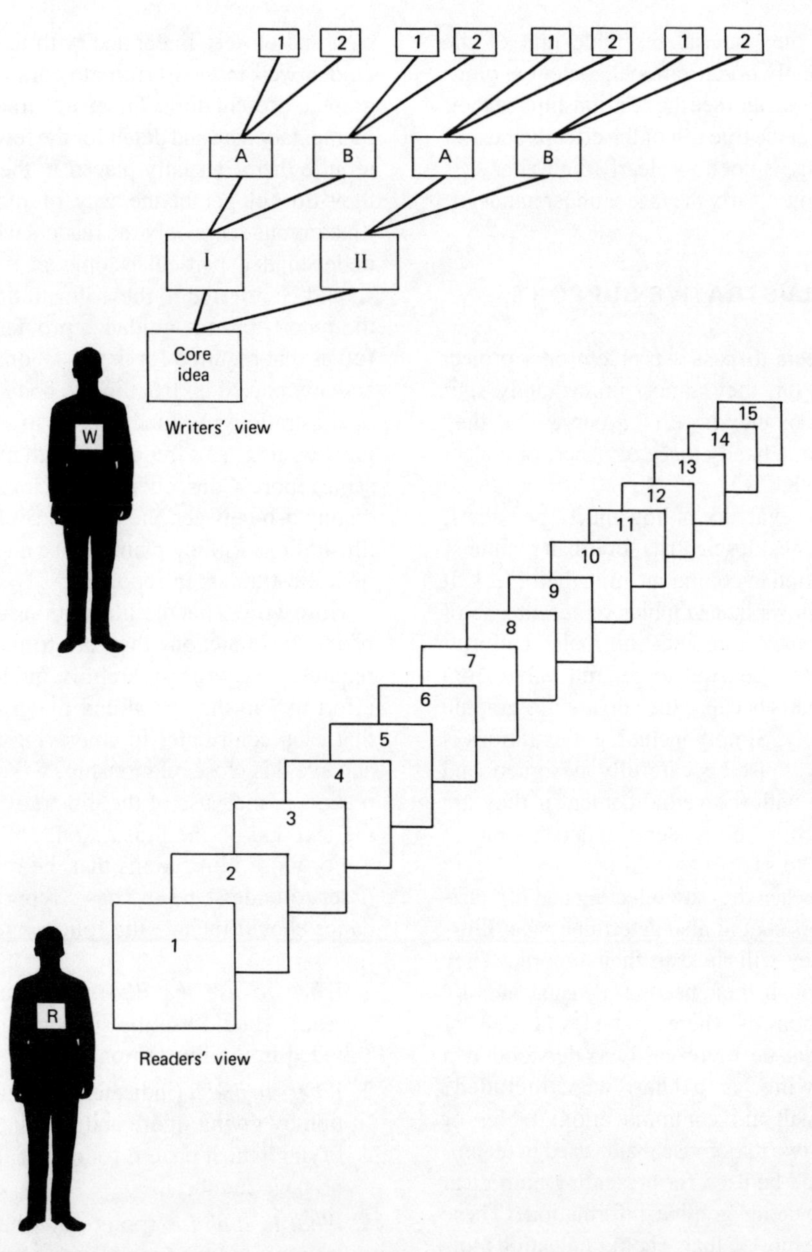

Figure 5. The writers have a sophisticated view of the material they organize. They see it both in terms of coordinate divisions and developmental directions. Readers see the information in a linear sequence—each idea comes after the other. Their view does not contain the sophistication of the writers. How does the writer provide the reader with the more sophisticated view?

the territory that is covered. By providing them with a perspective of the whole before presenting the first part, it prevents them from becoming confused or lost. Although adding this organizational content has required only a few words, it prepares readers for what follows. Such paragraphs do much to "cement" the sections and subsections of a report together; they make the whole a meaningful unit by providing readers with the relationships of the parts that bind the whole together.

In drafting their reports, whenever writers catch themselves placing one heading immediately after another, they should remind themselves to provide the reader with organizational content. It is not padding; it is as necessary to understanding as the informational content. It is the connective tissue that makes the reader's view of the material resemble that of the writer. Too frequently organizational content is omitted simply because writers see the relationships of their ideas so clearly. Yet the true test of the effectiveness of the piece of writing, is not how clearly the writer sees the material, but how clearly the reader understands it.

PROVIDING ILLUSTRATIVE SUPPORT

When two engineers discuss a problem or a project they are working on, they almost immediately start drawing sketches or the shape of a curve. All they require is a napkin, a handy piece of paper, or a clear spot on a tablecloth.

Such behavior is evidence of how much specialists, be they engineers, health scientists, or urban planners, depend on illustration in communicating their ideas. It most certainly follows that graphics represent one of their most important communication tools. Unfortunately, they all too often fail to get full value from illustrations in reports because they do not use graphic materials effectively. Simply including illustrations is not enough; they must be carefully designed and closely integrated with the verbal content if they are really to add much to the reader's understanding of what a writer has to say.

Consequently, when they are selecting and organizing content, writers should also determine what illustrative support they will need in their reports. They need to think through their needs for visual aids by asking such questions as where are tables needed, or would this material be more easily understood if a curve, or a drawing, or a chart were included? Whenever they will aid communication, tables or illustrations—the two major visual aids used in technical writing—should be used for presenting numerical values or for providing graphic information. These aids are among the most valuable communication tools writers have. But, if they are to be effective, writers must use them with purpose and, above all, must remember the needs of readers.

Two Functions in Reporting

Generally, illustrations serve one of two broad functions in technical writing. Some provide visual or statistical support that the readers need if they are to understand a report. Such illustrations, because they are necessary to understanding the basic ideas presented in the report, should be placed near the discussion and closely integrated with it. Readers must be guided well in their efforts to combine the verbal and graphic presentation. Other illustrations provide supplementary data and detail for the few readers who may require them. Usually placed in the appendix so that they do not get in the way of most readers, these illustrations serve only the readers who need or wish to understand a particular topic in depth. All readers should be referred to these illustrations in the body of the report, but the guidance provided need not be as full as that required for understanding the illustrations actually placed up front in the body itself. Both types of illustrations are fundamental to scientific and technical writing, and frequently both must be used in the same report. Consequently, it is important that writers distinguish between the functions of the two kinds of illustrations as they plan the use and location of their own illustrations in reports.

How writers handle illustrations depends on which of the two functions they are to play. The first type requires that writers carefully guide readers in their effort to join the verbal and the graphic materials so that each contributes in conveying the total idea. To achieve this close relationship, writers must guide the readers in their use of the illustrative material both in the text and on the illustration.

For those illustrations that the readers must use in order to understand the basic report, the textual guidance should include the following items:

1. *When to use the illustration* (indicate when the reader should use the supporting information provided in the illustration).
2. *Where to find it* (indicate the location of the illustration by giving figure and page number or by identifying which page it follows if the illustration has no page number).
3. *What to look for* (point out what information the reader is to look for and use as support for the ideas being discussed).
4. *What the significance of the material is* (state what relation the illustrative material has to the ideas being discussed in the text and what the importance of this relationship is).

Too often, writers fail to guide the reader well. The statement, "For results, see Figure 5," is typical of the guidance that the readers frequently get, but it is not enough. It provides only the first of the four items essential to proper integration. It fails to tell where to find the illustration, what to look for, or what significance the information in the illustration has. After being trapped a few times by such lack of direction, the reader quickly learns to skip the illustrations and "plow on" through the written discussion. There simply is no reason why readers should take time to work out what the writer should have provided.

How much easier it is for readers to combine the visual and the verbal materials when they receive guidance like that in the example below:

Local access to the reservoir shoreline is definitely limited (see Aerial Photo Mosaic, page 2). Although about 10 miles of state and county roads closely parallel the shoreline, access to the water from these routes is practical at only a few points. The steep, rugged character of the topography precludes increasing either the number of these points or early development of additional roads along the shoreline.

Here the writer effectively integrates illustration and text by providing the reader with the information necessary to intelligent use of the illustration. The first sentence tells *when* and *where*; the second, *what*; and the third, the *significance*.

Illustrations included to provide readers with depth of data should they require it, do not require the same amount of integration with the text. For such an illustration, it is usually sufficient to tell the reader:

1. *Where to find it.*
2. *What type of detail is to be found in the illustration.*

Consequently, references in the text can often be brief:

For manufacturing dimensions and tolerances, see Figure 17 in Appendix I, page 54.

This example meets the requirements for integration of illustrations included in an appendix. Readers can turn to these illustrations if their informational needs require doing so. They also know that they can skip such illustrations without any serious loss of content, without sacrificing their general understanding of the ideas being presented in the report.

If writers are really to guide readers in integrating visual and verbal content, however, they must do more than provide guidance in the text of the report. They must also see to it that the eye of the reader is guided on the illustration itself. (See Figures 6 and 7, pp. 44 and 45.) Here their goal is to help readers in quickly and easily discovering the significant information an illustration is providing. The eye of the reader can be directed by a variety of very useful illustrating techniques, including the use of callouts and arrows; the use of distinguishing markings, such as cross-hatching and dot patterns; and the simplicity of the design of the illustration itself. When these techniques are used effectively, readers can easily recognize and quickly assimilate the content presented by an illustration.

When to Use

Two views are commonly held of the use of illustrations: (1) that illustrations are provided to *supplement* the verbal presentation of information or (2) that illustrations are presented to *supplant* verbal explanation. The most realistic position actually lies somewhere between these two points of view. Certainly, an illustration can substantially *reduce* the number of words necessary to convey an idea, but it cannot completely eliminate the need for some words, because the illustration and the verbal text must work together in conveying the total idea. At best, an illustration assists readers in understanding an idea more fully and more easily than they would without it. In a sense, illustrations are to the report as visual aids are to a talk, and visual aids, regardless of their excellence, do not replace the speaker. In fact, illustrations without explanation frequently raise more questions than they answer. Consequently, rather than helping readers to understand, unexplained illustrations block communication by confusing issues and raising unnecessary questions.

This *concept of support* carries with it a number of serious implications for writers. For example, an illustration must contribute content to the report and be integrated with the discussion so that the reader can join the illustrative support with the ideas being discussed. At the same time, if writers are to use illustrations effectively, they need to know how illustrations contribute content and meaning to a report. Most often, illustrations can provide

1. *Detail difficult to describe verbally;* for example, the crystalline structure of a metal, fatigue failure of metal, corrosion or wear, etc. Figure 8 on page 46 is a revelation of detail that words alone probably could never represent adequately.
2. *Overall relationship of detail* in a little space; for example, an assembly drawing, a flow chart for a chemical process, a represenation of a back-up pool behind a proposed dam, curves, numerical data in tables, and so on. The map in Figure 9, page 47, clearly and concisely reveals a complex set of geo-

Figure 1 - View looking up Mill Creek showing flood channel at eastern Portion of Walla Walla. Upper end of concrete flood channel is in center foreground and wire bound groins of improved channel in center of photo.

Figure 6. The writer tried to guide the reader by including the directions in the caption, but the illustration is still difficult to use. Visual guidance is needed as in Figure 7. (Photo courtesy U.S. Corps of Engineers.)

graphical details in relationship to the National Reactor Test Station in southeastern Idaho.

3. *Emphasis* by giving ideas, problems, or situations visual impact; for example a photograph of a flooded valley, a before-and-after view showing destruction or improvement, or a table of substantial cost savings. What better way to emphasize the writer's point about the effects of erosion than the photograph in Figure 10, page 48?

Writers should be constantly alert to identifying when illustrations might be most helpful in communicating ideas to readers. A helpful practice is to note on the outline itself what illustrations are needed and where. Then, before sitting down to write the first draft, it is equally helpful to gather together the illustrative materials so that they are at hand to supplement the outline and guide the writer during writing. Frequently, the need for an illustration may not become clear until a subdivision of the outline has taken shape in words and sentences, and the alert writer will keep an eye open for this emerging need. Doing so will reduce the number of occasions on which writers fail to include a helpful illustration, as in the following example.

Faced with describing some geographical relationships, an author did not provide an illustration with this paragraph:

Youngs Creek originates on the western slopes of the Bear Mountains about 21 miles east of the city of Three Rivers. The stream flows in a westerly direction through the center of the city to a point about 6 miles west of the city, where it enters the north fork of the Big Sandy River. The chief contributary to Youngs Creek is Bear Creek, which enters Youngs Creek about 9 miles above the city of Three Rivers. Yellow Hand and Rocky creeks are small delta avulsion branches of Youngs Creek that pass through Three Rivers and join the north fork of the Big Sandy River upstream from the mouth of Youngs Creek. These creeks carry portions of the

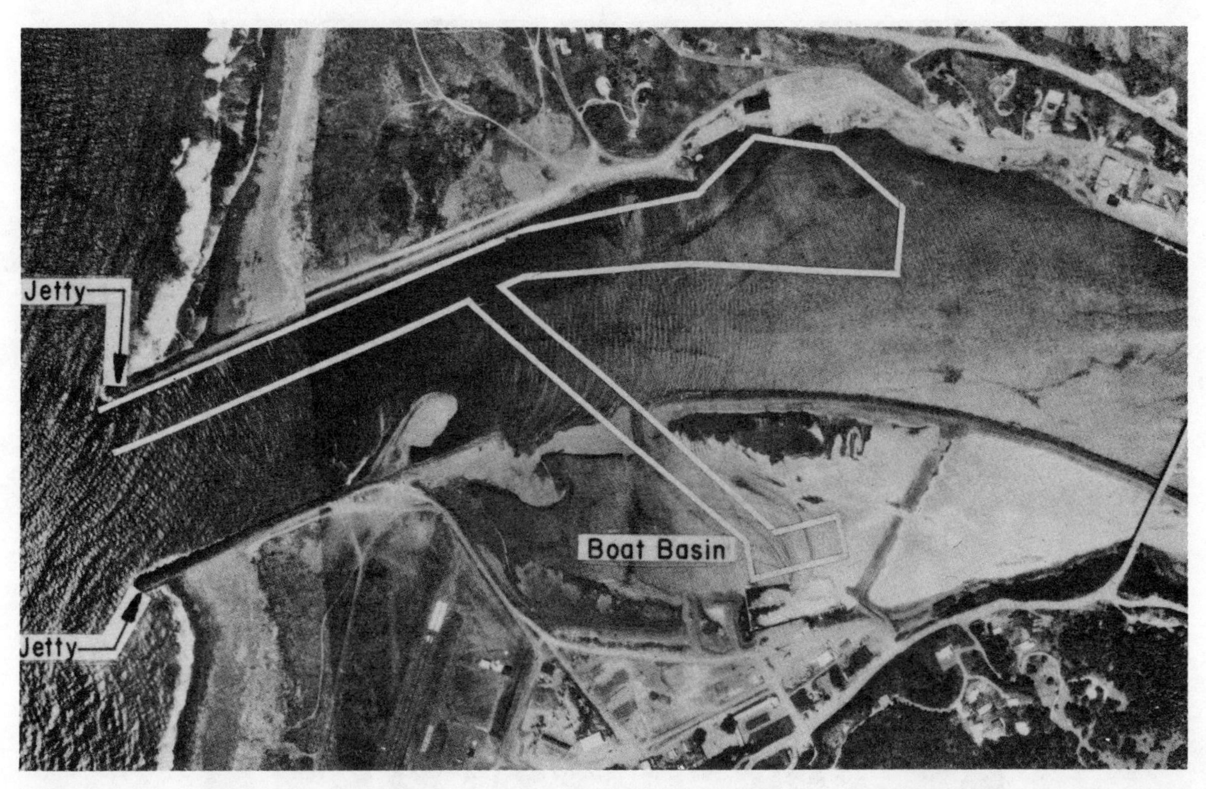

Figure 7. Here the callouts and arrows help the reader find the items quickly and easily. It would help, perhaps, if the callouts were larger. (Photo courtesy U.S. Corps of Engineers.)

flood overflows from Youngs Creek and are also part of the transportation channels that carry irrigation water which is diverted from Youngs Creek at the division works at the existing floor control project.

In this case, the relationships are not highly complicated; yet how much clearer they would have been for the reader had the writer included a simple sketch. To understand so simple a geographical description as this one, readers must create their own maps, sketching the relationships mentally or on a piece of paper. The real question, of course, is why they should find it necessary to do what the writer should have done?

Designing and Locating Illustrations

In designing tables and illustrations, writers should keep in mind that "simplicity is the key to effectiveness." It is important to realize also that working diagrams and drawings are frequently not effective when used as illustrations. Because illustrations are intended to illustrate a particular point, all the material included should be relevant to that point; all else should be omitted. Working drawings usually contain much more information than the reader needs; consequently, the important material is obscured by a mass of unnecessary and distracting detail. Writers, then, should design illustrations that include all of the information necessary to support the text—but no more! Be it a table, a curve, a flowchart, a schematic, an exploded view, or a photograph—simple, direct design is the key to effectiveness. (See Figure 11, p. 48.)

If writers have professional illustrators to assist them in the preparation of their illustrations, they should discuss with the illustrators what it is they want to illustrate, what items of information must be included, and how the various items relate to one another. Writers should, however, allow the illustrators to use their own imaginations and their own professional training and experience in determining how the material might best be presented graphically.

The use of color should be avoided in illustrations that are to be photographed and reproduced in black and white, for reds come out very black, and blues, greens, and oranges tend to fade. Commercial "stick-down materials," such as zipatone, provide cross-hatch, callout, and letter patterns that make pre-

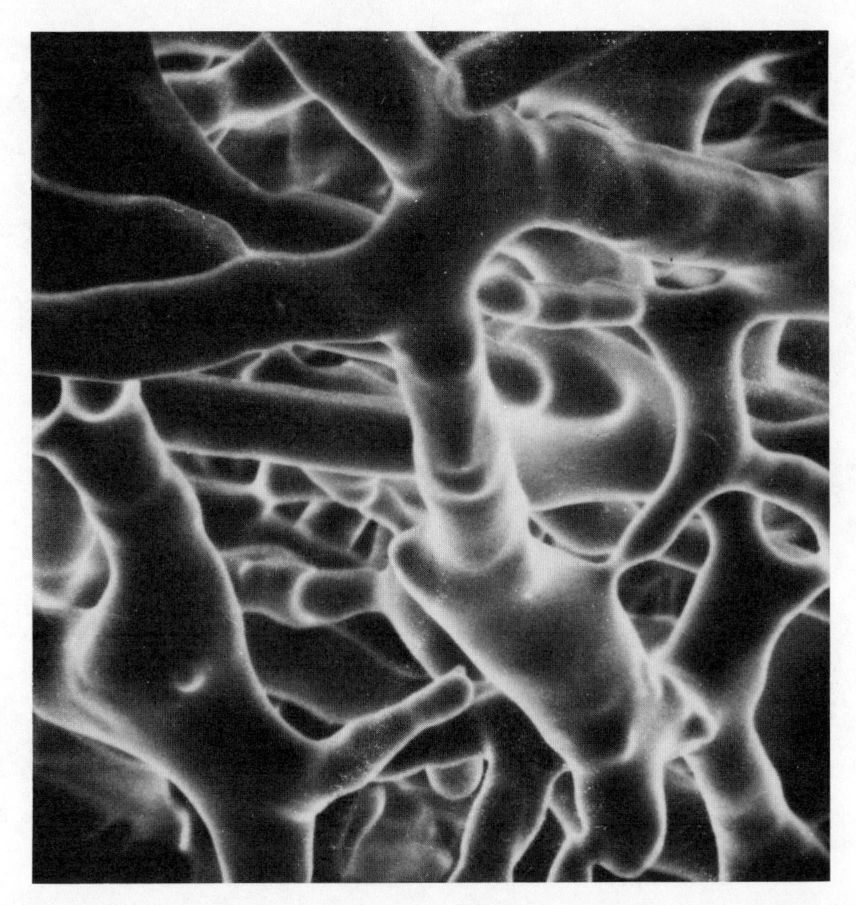

Figure 8. Illustration showing detail difficult to describe verbally. How accurately could this be put into words? (Photo courtesy National Aeronautics and Space Administration—Ames Research Center.)

paring illustrations quick and effective by eliminating much of the inking. It is important to remember, however, that distinctive patterns must remain distinctive after reduction whenever large-scale illustrations are prepared to be photographically reduced before being put into a report. When illustrations are sent to the photographic laboratory, the writer should mark the top or bottom so that the orientation of the picture to the page is easily recognized in the laboratory and should clearly indicate what the size of the final illustration is to be.

Every figure (illustration) should have a number and a caption. Figures in a report are usually numbered consecutively, and the figure numbers and the captions are placed below the illustrations. If the bulk of detail is toward the bottom of an illustration, sometimes better balance can be achieved by placing the caption in a box within the upper part of the illustration itself.

Tables are also numbered consecutively, but their numbers and titles are usually placed above the table, rather than below as with illustrations. It is important in the preparation of tables to remember that all factors that affect data must be indicated so that the reader can fully understand the material being presented. If the factors consistently apply to the entire table, they can be listed following the table's title. If a factor applies to a specific column, it probably should be included in the column title or column head. Factors having limited application can also be presented in footnotes placed below the table.

Generally, those illustrations and tables that are necessary for full understanding of a report's message are best placed in the report as close to the verbal discussion as possible, either on pages with text or between pages of text. Those that provide an in-depth view for the reader who wants to pursue a particular topic in greater detail are usually best grouped in the appendix of the report, out of the way of those readers

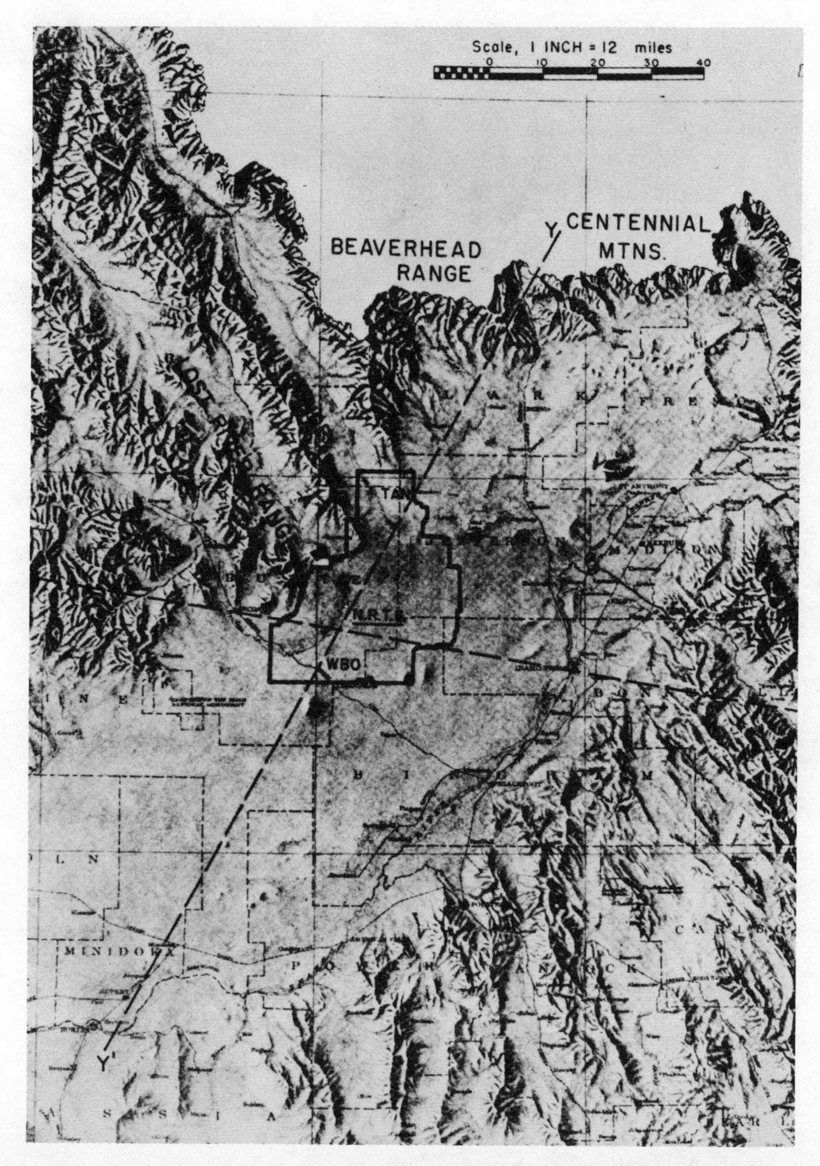

Figure 9. A map showing the location of the National Reactor Test Station. A great amount of topographical detail is presented in a highly concise and related way. How much writing would be required to describe (1) the shape of the station and (2) its relationship to the various topographical features shown here? (Photo courtesy U.S. Atomic Energy Commission.)

who do not need the in-depth view. At no time should an illustration precede the verbal discussion. It should either be presented with or follow such discussion. Illustrations placed on a page with text should be readable without having to turn the report sideways. Large illustrations are best placed on a separate page with the figure number and caption at the bottom of the illustration, whether this be at the bottom or the outside margin of the page. Whenever possible, tables should be held to one page rather than be continued over to a second page, for carrying information from one page to another is difficult for the reader.

In using illustrations, then, writers should seek simplicity of design and maximum contribution of meaning. They should present each illustration so that the reader may use it easily. They should integrate each illustration well with the verbal text which calls the reader's attention to it and should provide callouts on the illustration itself to guide the reader's eye. Above all else, however, illustrations should contribute con-

Figure 10. This picture gives visual impact to the idea that concentrated runoff can cause severe road damage. (Photo courtesy State of Montana Department of Conservation and Natural Resources.)

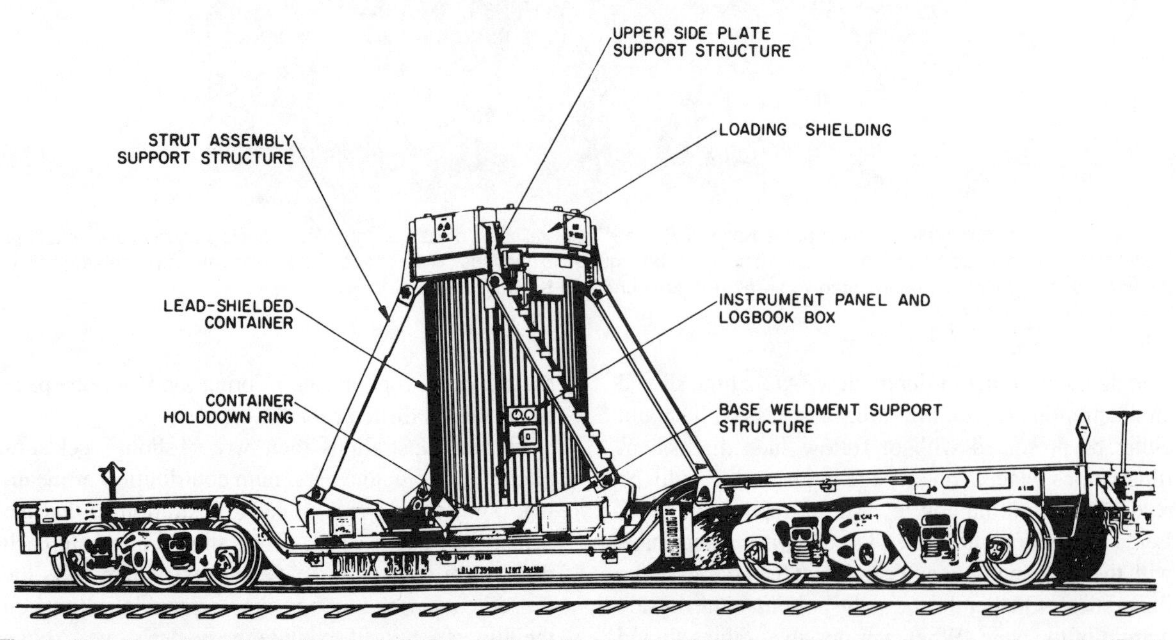

Figure 11. An excellent illustration combining simplicity of design with effective emphasis and clean use of callouts. (Photo courtesy U.S. Energy Research and Development Administration.)

tent to the report and support the verbal discussion. Although illustrations should be complete in themselves, they do not really stand alone.

Most often, poor illustrations result from the failure of the writer:

1. To design an illustration to support a specific idea—settling for the conveniently available working drawing, with all of its clutter, rather than developing the effective illustration.
2. To direct a reader in the use of the illustration —failing to integrate the illustration with the text and to direct the reader's attention on the illustration itself.
3. To anticipate the effect of reduction on the readability and clarity of the call-outs, lines, data points, cross-hatched areas, and other elements of the illustration.
4. To consider the capabilities and limitations of specific photographic processes in reproducing the contrasts of several colors in black and white.

To be effective illustrations must

1. Contribute useful content.
2. Be properly integrated with the text.
3. Be located at a point of greatest convenience for the reader if necessary to his understanding or placed in the appendix if provided for in-depth view for those readers who need such a view.
4. Be clear and self-explanatory.
5. Be accurate and not distort the facts.
6. Be clear and as simple as possible and yet convey the information.
7. Be well designed so that the content and the relationships are clearly grasped and easily understood.

DETERMINING FORM AND LAYOUT

In determining the form and layout of their reports, writers are actually confronting choices in three separate and distinguishable areas: external form, the internal layout of the text's headings and subheadings, and page layout. Choices in each area are influenced by different factors; yet in combination they determine to a large degree the physical appearance a report takes. Very often, the form is dictated by company policy, by tradition, or by the request or specifications of a customer. When this is not the case, however, writers must decide on a report form for themselves, and it becomes important to know what factors influence the selection of a report form. They must also decide what internal layout will best represent the organization of a report's contents. They must choose the headings and subheadings that will serve best to label the divisions and subdivisions of their material and decide on the physical arrangement of their labels that will show clearly the organizational relationships. Finally, writers must determine which content in a report requires the special emphasis that effective page layout can provide.

Unfortunately, many writers fail to distinguish between these tasks of determining report form, internal layout, and page layout. As a result, they can make poor choices in each area for the wrong reasons, and the physical appearance and usefulness of a report can leave something to be desired. If their choices are to be effective, writers should recognize that decisions in each area must arise from different communication factors.

1. Choice of report form depends on the circulation of a report, the relationship to be established between writer and reader, the use to be made of the report, and the amount of material the report will contain.
2. Choice of internal layout depends on the organizational relationships of the units of content in a report.
3. Choice of page layout depends on the desired emphasis and unity of the material presented.

Much of the misunderstanding about matters of report form and layout is kept alive by widely used prescriptions in industry and government which include too much in the name of report form. Many prescriptions not only include instructions about such items as the cover, the title page, and the table of contents but also specify what the actual headings and subheadings of a report should be. Thus they do more than provide rules or guidelines for what a report should look like; they dictate what its content and organization should be. Such prescriptions, whether set down in writing or established by tradition, fail to recognize that standardizing forms may not prove troublesome but that standardizing content and organization creates real problems for writers who would report effectively on their material.

Choosing Report Form

Report form is determined by the amount and complexity of the material to be included, the desired writer-reader relationship, and the use and circulation of a report.

Selecting report form is really a matter of choosing a container for the information to be provided the reader. Form is packaging. The more information to be provided, the larger the container must be, or to be more specific, the more structural devices for handling content that the report form must have. Report form, as any container, must meet its "load" requirements and the demands of its use environment if it is to be effective. A memo or a letter, for example, is a container designed for small "loads," or messages. Neither requires a table of contents and, most often, not even a summary. The formal report, on the other hand, is a container designed for relatively long and complex messages; consequently, it has a number of additional structural devices to control its greater content: title pages, tables of content, lists of illustrations, summaries, appendices, and the like.

Report form also aids in establishing a specific writer-reader relationship. A memo can establish an informal relationship. A letter addressed to a specific person or limited group of people can create a direct, personal relationship between writer and readers, much as the memo does, but one that is not necessarily so informal. An article or a formal report is addressed to many readers and is thus more formal and more impersonal than the memo or letter. The tone appropriate to a given reporting situation, then, should be supported by the form into which a report is placed.

The great variety of writing forms found in industry and government has resulted largely from the many uses made of reports and the circulation given them. Actually, most of these forms can be seen as belonging in two broad categories: correspondence forms and report forms. Many persons, perhaps, are not accustomed to thinking of business letters or memos as reporting forms; yet the fact is that numerous scientists and engineers have put brief reports into letters and memos, as well as using them in the more conventional way.

Correspondence Forms

The two primary correspondence forms are letters and memos. Each is used for a particular kind of circulation. The letter form may be used for correspondence circulated within an organization but is almost always used for correspondence going outside of it. A memo, on the other hand, is an abbreviated or informal correspondence form most often restricted to correspondence within a company or an agency. It is seldom used for "outside" correspondence. (See Figure 12, p. 51.)

Letters and memos are particularly marked by two characteristics—first, they provide for a personal relationship between writer and reader, frequently on a one-to-one basis, and, second, they are usually short and confined to one subject.

For example, to inform a supervisor of the need for additional personnel, the completion of a research project, and the transfer of equipment to another department, it is usual to write a separate memo on each subject, rather than take up all three in one piece of correspondence. At first glance, this practice may seem foolish and wasteful. Why write three memos, when one can do the job? Of course, the answer is that forwarding one memo to three different places—the Personnel Department, the Research Director, and the Equipment Control Department—can present difficulties. If all three subjects were included in one piece of correspondence, either the supervisor would have to write directly to each of the three offices concerned, or the personnel of each would be forced to read material in the original of which only one-third relates to their responsibilities. Moreover, one memo cannot readily be filed in three different places; thus, the records in two places would be incomplete. Each desired action requires a different circulation, a different company routing; consequently, correspondence is confined to one topic, to one problem.

In industry and government everyone, including the scientist and engineer, can waste valuable time in reading disorganized, confusing correspondence. Consequently, those writers are appreciated and rewarded who get to the point quickly and provide a complete, yet brief and understandable message. This combination of being concise on the one hand and including all of the essential information on the other is the mark of excellence in correspondence. Its value to a company, as well as to the individual, cannot be overestimated.

It is important also when writing correspondence, whether it be a letter or a memo, that the tone of the message be reasonably warm and friendly, rather than cold and impersonal or even unconcerned. The best guides for writing correspondence are real consideration for readers and their problems and a sincere wish to assist them.

Report Forms

The report forms commonly employed in industry and government fall into one of four classes, and the choice

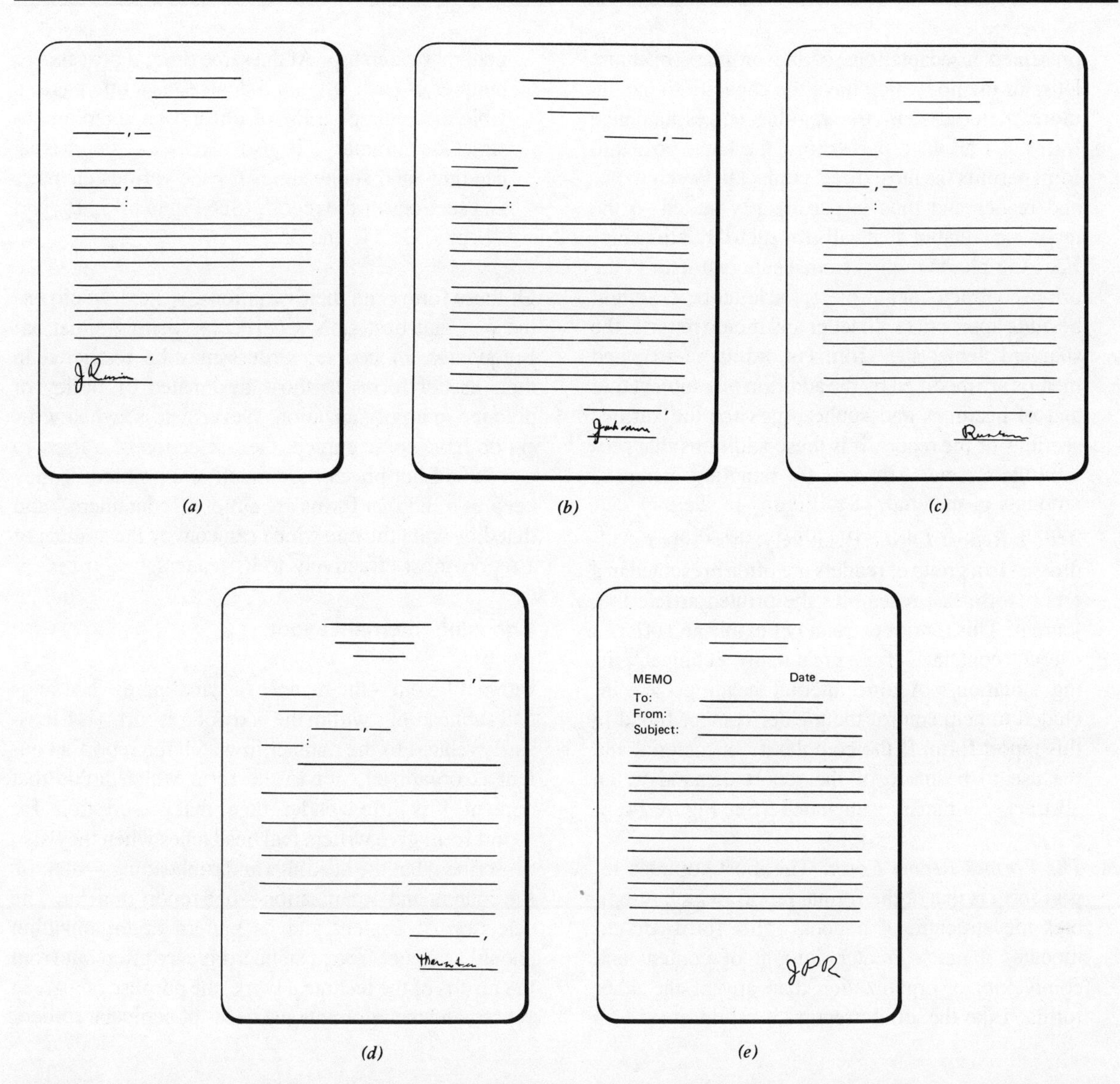

Figure 12. Correspondence forms: letter forms: *(a)* full block, using left-hand margin, newest form; *(b)* block form, most widely used letter form; *(c)* semiblock form, used widely but slightly conservative; *(d)* indented form, used for double-spaced letters and short messages; memo form: *(e)* a widely used, standard form.

among these forms depends on the use to be made of a writer's report, as well as on the length, complexity, and the nature of the material to be presented. Writers should remember that they really are choosing "containers" for their information when they decide which of the following forms is suitable for their reports.

1. *Form Reports*. Form reports are most commonly used for routine reporting situations in which the same types of information are called for periodically. A good example is the federal income tax form. Generally, the form report is used for maintaining a record and checking on progress. Frequently, it is designed to provide specific information for a larger composite report. Progress reports from various groups, for example, are often used as the basis for a manager's report on the work of an entire department or division, or production reports from the different production lines in a plant can provide the content for a report on the plant's total output. The effectiveness of form reports depends on how well the forms have been designed for obtaining the specific information needed.
2. *Correspondence Report Forms*. Reports are often

presented in adaptations of the ordinary business letter or memo, which have the capacity to handle more material than the regular correspondence forms. Of all the report forms, the letter or memo form permits the most direct contact between writer and reader and thus is particularly suited to the reporting situation that calls for such a relationship. For example, in reports from a subcontractor to the prime contractor of a project, the letter report might be quite appropriate. In letter and memo reports, the standard elements and forms of ordinary letters and memos are modified by the addition of a subject title and of headings and subheadings for the various sections of the report; it is these additions that provide the structural devices for handling increased amounts of material. (See Figure 13, *below*).

3. *Article Report Form*. Relatively short reports addressed to a group of readers are often presented in a report form that resembles the printed article in a journal. This form is extremely flexible and offers a useful "container" for a great many technical writing situations. Again, internal headings are included to help control the greater content found in this report form. If the complexity of material and the use to be made of the report demand it, an abstract or summary is included. (See Figure 14, p. 53.)
4. *The Formal Report Form*. The most complex report form is that of the formal report, which resembles the structure of a book. This form accommodates a much greater amount of content and complexity of organization than any of the other forms. Like the article report, it is addressed to a general readership. At the same time, it possesses a number of such structural devices as a title page, a table of contents, a list of illustrations, and an abstract or summary. It also makes use of internal headings and subheadings for the various chapters and sections of the report. (See Figures 15, 16, and 17, pp. 53, 54, and 55.)

Of these forms and their variations, writers should use the one that best suits a particular writing situation. Frequently, of course, writers may be restricted in their use of forms to those designated by policy or practice in an organization. Nevertheless, when writers do have some choice, their selection of a form to use should not present too much of a problem if they keep in mind that forms are simply "containers" and that they want the one which can convey the content of a report most effectively to its readers.

Choosing Internal Layout

Internal layout—the manner of handling the headings and subheadings within the body of a report—is necessarily related to the manner in which the report's content is organized, not to the form which holds that content. It is little wonder, then, that prescriptions for report form give writers real headaches when they also prescribe what the headings and subheadings—that is, the content and organization—of a report shall be. The selection of content and its pattern of organization should arise not from a standard prescription but from the results of the technical work, the purpose of a given report, and the informational needs of particular readers.

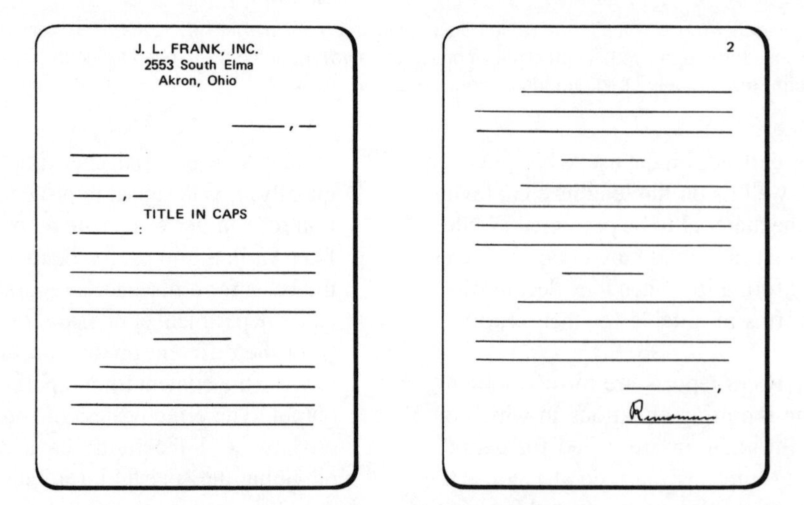
J. L. FRANK, INC.
2553 South Elma
Akron, Ohio

TITLE IN CAPS

2

Figure 13. Letter report form.

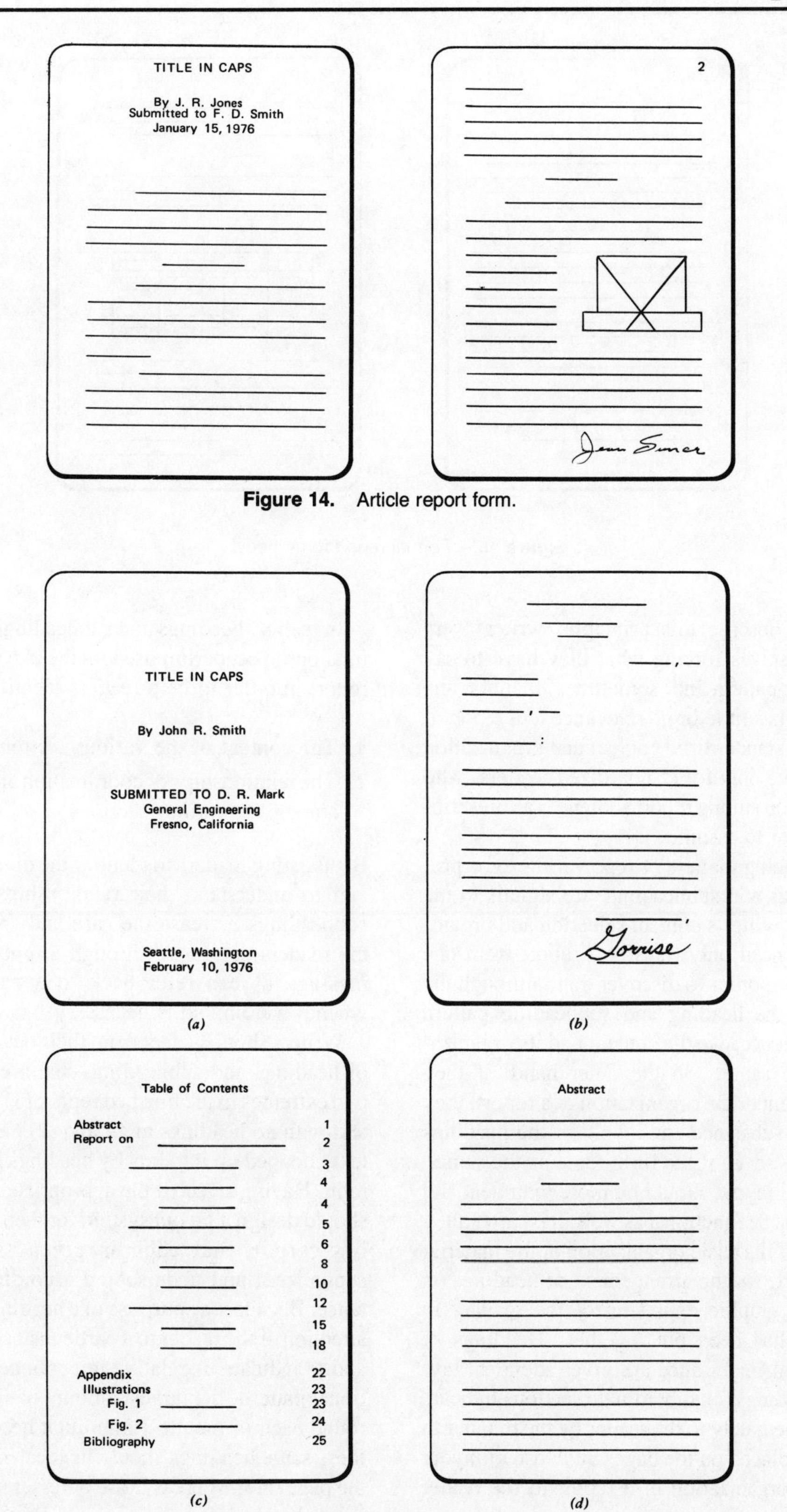

Figure 15. Formal report form: preliminary parts: *(a)* title page, *(b)* letter of transmittal, *(c)* table of contents, and *(d)* abstract or summary.

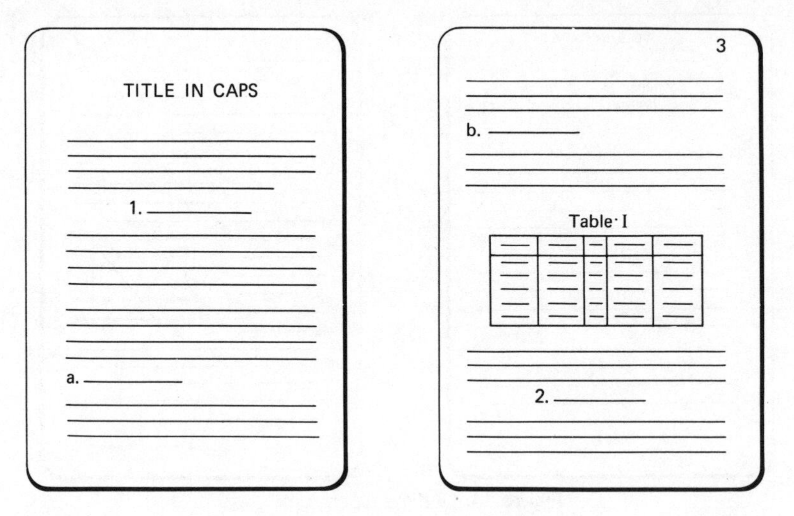

Figure 16. Formal report form: body.

If they cannot observe this principle, writers very often find themselves forcing what they have to say into an ill-fitting pattern and, sometimes, manufacturing content that has little or no relevance to a report at all. In any case, standardized content and organization are fully effective only for standardized readers, who are not as common among report audiences as prescription writers seem to assume.

Rather than being matters of report form to be prescribed, headings and subheadings are signals to the reader about the writer's content selection and organization. Writers need only transfer a report from one form to another in order to discover that, although the form changes, the heading and subheading pattern remains constant because the content and its organization remain the same. On the other hand, if they change the content or the organization of a report, they must revise the substance and layout of the headings and subheadings, even if the form remains the same.

Good internal layout should be more than neat and attractive; it must be functional as well. It is, after all, a representation of the basic organization of the material within the report, for the arrangement of headings on each page is a graphic expresion of the manner in which content has been put together. Headings of sections of equal importance are given identical layout, and subheadings for subordinate sections indicate their status immediately to the reader by the manner in which they are placed on the page. Such use of layout can reveal the organization of a report to the reader very quickly.

In reality, headings and subheadings are parts of the final outline superimposed on the expanded text of the report in order to help readers identify

1. The content of the various sections of the report.
2. The relationships of coordination and subordination among the various sections.

By assisting readers to identify the divisions of content and to understand their relationships, headings and subheadings increase the ease and speed with which the readers can move through a report and grasp its message or can refer back to a particular section whenever doing so is necessary.

Writers should select from their outlines the number of headings and subheadings that are likely to avoid two extremes in the finished report: (1) several pages of text with no headings at all and (2) page after page of text chopped up into bits by headings for every paragraph. Having selected the appropriate number, writers should design a layout system for their headings, making certain that each is given space, located, capitalized, and underscored according to its importance. Because the purpose of a heading is to attract the attention of the reader to a particular section, to label it, and to indicate its relative importance in an organizational pattern, the largest amount of space should surround each of the most important headings. Probably these same headings should be located in the center of the page line, whereas those of lesser rank can often be located at the left-hand margin. Very frequently, main

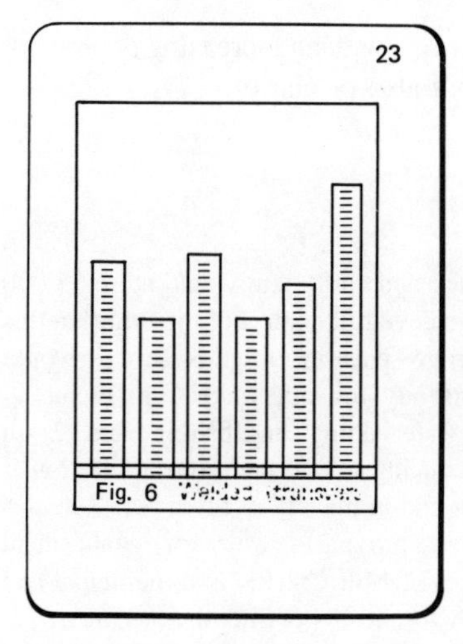

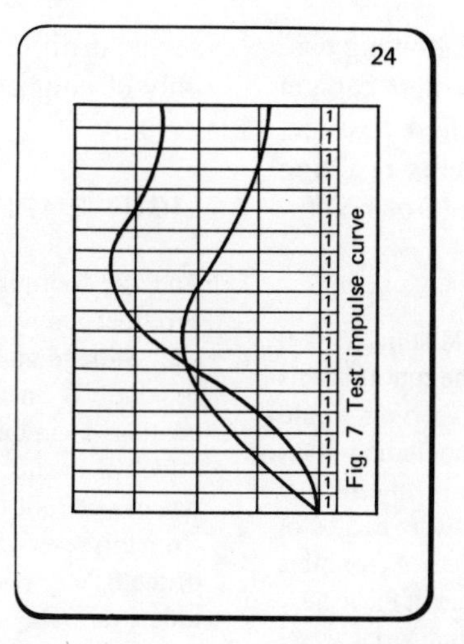

Figure 17. Formal report form: supplemental parts.

headings are fully capitalized to attract attention, and, at times, they are underscored for the same reason.

These important considerations in designing a layout system have been restricted, of course, to the devices of layout that are available on any typewriter. Although many reports are set in type and thus their writers have available the much larger number of layout potentialities a variety of typefaces makes possible, it is fair to say that the reproduction of most reports today is still tied in some fashion to the typewriter. This fact, however, does not mean that effective layout is impossible, or even difficult, to obtain. Writers can achieve this objective by making use of some rather basic principles—that, for example, the more space a heading has around it, the more emphasis it gets, the more important it looks; or that a centered heading looks more important than one in the left-hand margin. It is also worth remembering to use several layout devices—space, location, capitalization, and underscoring—in a variety of combinations, rather than to use just one or two.

The following list indicates the relative emphasis an ordinary typewriter can give to headings; regardless of their position on a page:

1. CAPITALIZED, UNDERSCORED.
2. CAPITALIZED.
3. Capitals and Lowercase, Underscored.
4. Capitals and Lowercase.

Combining these varieties of emphasis with a variety of locations can produce, among others, the two commonly used layout systems shown in Figure 18 on page 56.

Page Layout

The use of the term *page layout* here refers to the way in which the words, phrases, and sentences of a report's text are actually put down on a piece of paper. By means of the contrast between white space and black type, effective page layout can provide emphasis and unity for ideas being presented. It can also contribute significantly to briefness and clarity of presentation.

Even when reports are typed, writers have a number of techniques available for their use in emphasizing ideas. Some of these are

1. Single spacing versus double spacing.
2. Providing extra space at the left-hand margin by underhanging the copy.
3. Enclosing material in a box or presenting it in columns.
4. Underlining key terms and ideas.
5. Numbering parallel ideas in a sentence or paragraph.
6. Putting ideas into a list and marking the beginning of each item with numbers, dots, hyphens, dashes, or other symbols.

Used wisely, such techniques permit writers to give varying degrees of visual impact and hence emphasis to different parts of their content. The following passages demonstrate some of the effects writers can get with variations in page layout. In the first passage, taken from a technical report, the benefits resulting from the project being described get little or no emphasis.

The 272-foot-high rock and earth dam would store 53,000 acre-feet and would achieve full control of the runoff at this site. Storage and storage space in Gorge Reservoir would provide an opportunity for lake recreation, including boating, swimming, water skiing, and fishing, and an effective means of reducing flood conditions in the lower reaches of Mill Creek and in the city of Sheridan. Releases from this reservoir would provide water for enhancement of the fishery resources of Mill Creek, a municipal and industrial water supply for the city of Sheridan, an irrigation water supply to 15,000 acres of good quality lands, and a substantial contribution toward meeting future minimum water quality flows in the South Yamhill and Yamhill Rivers.

Interestingly enough, however, one of the purposes of the report was to convince persons in decision-making positions that the project should be undertaken and, at the same time, to gain public support for it. Therefore, it seems reasonable to believe that some emphasis on the benefits would have served the report's purpose very well. Just how much emphasis they should have received in this particular paragraph is a question that cannot be answered in the absence of the remainder of the report, but the important point is that the writer had some choices. The next three passages demonstrate how the same material may be presented in different ways, with an increasing degree not only of emphasis but also of clarity.

LIGHT EMPHASIS

The 272-foot-high rock and earth dam would store 53,000 acre-feet and would achieve full control of the runoff at this site. Storage and storage space in Gorge Reservoir would provide (1) an opportunity for lake recreation, including boating, swimming, water skiing, and fishing, and (2) an effective means of reducing flood conditions in the lower reaches of Mill Creek and in the city of Sheridan. Releases from this reservoir would provide (1) water for enhancement of the fishery resources of Mill Creek, (2) a municipal and industrial water supply for the city of Sheridan, (3) an irrigation water supply to 15,000 acres of good quality lands, and (4) a substantial contribution toward meeting future minimum water quality flows in the South Yamhill and Yamhill Rivers.

MODERATE EMPHASIS

The 272-foot-high rock and earth dam would store 53,000 acre-feet and would achieve full control of the runoff at this site. Storage and storage space in Gorge Reservoir would provide (1) an opportunity for lake recreation, including boating, swimming, water skiing, and fishing, and (2) an effective means of reducing flood conditions in the lower reaches of Mill Creek and in the city of Sheridan. Releases from this reservoir would provide (1) water for enhancement of the fishery resources of Mill Creek, (2) a municipal and industrial water supply for the city of Sherican, (3) an irrigation water supply to 15,000 acres of good quality lands, and

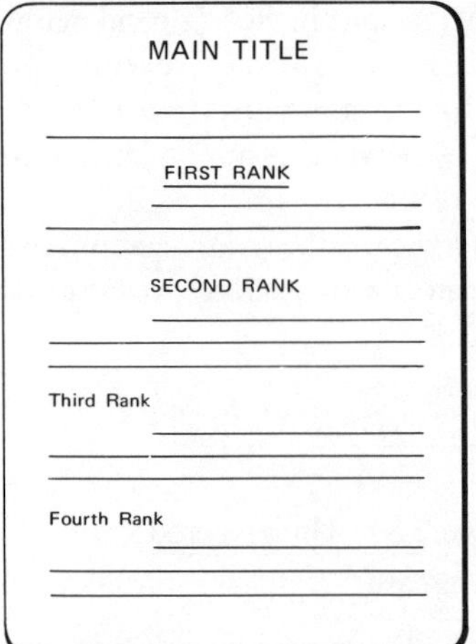

Figure 18. Use of headings to reveal organizational relationship of the various sections.

(4) a substantial contribution toward meeting future minimum water quality flows in the South Yamhill Rivers.

HEAVY EMPHASIS

The 272-foot-high rock and earth dam would store 53,000 acre-feet and would achieve full control of the runoff at this site.

The Gorge Reservoir would provide two kinds of benefits: those from reservoir storage and those from reservoir releases.

- *Storage Benefits*
  1. An opportunity for lake recreation, including boating, swimming, water skiing, and fishing.
  2. An effective means of reducing flood conditions in the lower reaches of Mill Creek and in the city of Sheridan.

- *Release Benefits*
  1. Water for enhancement of the fishery resources of Mill Creek.
  2. A municipal and industrial water supply for the city of Sheridan.
  3. An irrigation water supply to 15,000 acres of good quality lands.
  4. A substantial contribution toward meeting future minimum water quality flows in the South Yamhill and Yamhill Rivers.

Page layout can be a useful tool for the technical writers who remain alert to its possibilities. Indeed, this observation holds true also for internal layout and form. Once writers understand the difference between external form and internal layout and identify the factors that influence the use and choice of each, they can make both of these report elements work for them. They can effectively package their ideas and emphasize the relationships they see in the material they present. Form and layout, then, become positive aids, rather than troublesome handicaps to good reporting. They become tools rather than handcuffs.

EXERCISES AND DISCUSSION TOPICS

1. Develop the detail for the overall pattern of presentation that you have selected for your report, and prepare a formal outline to serve as your writing guide. Identify the various orders that you have used, including both the overall pattern and the internal patterns for the various sections of your outline.
2. Examine a report or article. Identify the overall pattern and the various orders used in the internal sections of the report or article. How many different orders were used? What are they?
3. Write an introduction for your report that reveals its organization for the reader. Write two internal introductions to prepare the reader for the subdivisions of two of your main sections.
4. Prepare two illustrations that you are going to use in your report. Include alternative designs available to you for the illustrations, and explain why you chose the particular design you did. For *each* illustration, write the reference that you will include in your report in order to guide the readers in their use of it.
5. Using principles of effective layout, show how you would retype the following paragraphs to clarify the actions for each year of the construction schedule.

Construction Schedule

A three-year construction period, preceded by a year of advance planning and followed by one year of testing, would be required.

Activities scheduled the first year are completion of the advance planning report, negotiation of the repayment contract, drilling an exploratory well at each site, and establishing the Bureau field office.

During the second year, field data for use in preparing designs and specifications for the wells and lateral systems would be gathered, and a contract would be awarded for drilling the wells.

The well drilling would be completed the third year, and a contract to construct the lateral systems would be let. Construction would start on the operation and maintenance buildings.

The fourth year, the pumps and motors would be installed, the Twin Lakes access areas would be constructed, other work in progress would be completed, the first delivery of water would be made, and testing and minor rectification work would be started. Testing and minor rectification would be completed early in the fifth fiscal or final construction year.

6. Develop a layout pattern for the headings and subheadings of your report. Describe the layout for each rank of heading and explain how it physically represents the relative importance of that rank of heading.

CHAPTER 6

Applying the Design

Once a design is complete, writers must get down to applying it, that is, to actually writing the letter, memo, or report. The writing activity involves three steps: fully stating a content in sentences and paragraphs; revising this first draft until the result meets the requirements of purpose, use, and audience; and then preparing the final copy. When the final copy has been checked, approved, and reproduced, the writing product is finished—it is ready for distribution, ready to serve its readers.

No doubt, all writers wish they could proceed from the outlines to the finished product as easily and simply as that summary does. Unfortunately, writing seldom proves to be easy and simple. It requires hard, often tedious labor, and it presents yet a different set of challenges. At the beginning comes the difficulty of getting started on writing at all. Then follows the wrestling match with the writing activity itself, a contest in which writers can fare badly if they do not understand its nature. Finally, particularly during revision, writers confront the problems of attaining an effective tone and style.

PROBLEMS OF GETTING STARTED

For many writers getting started on the first draft is a major problem. Given an opportunity, they will put off writing as long as they can. In fact, some writers do not even need an opportunity; they will create one. In writing, procrastination is more the rule than the exception, for almost any excuse serves to postpone writing.

What can be done about this problem? Little, except to face up to the tendency to put off writing and to resist it. Sometimes, publicly establishing deadlines with a supervisor helps, provided that the writer has developed a healthy respect for meeting them. Firmly planning and scheduling time for writing is another defense against procrastination. And, of course, supervisors can discourage procrastination by holding writers to sensible deadlines.

Yet, as Dr. Henrietta Tichy suggests in her excellent book, *Effective Writing,* John Wiley & Sons, New York, 1966, the difficulty in getting started is most often the result of not being prepared, of too many interruptions, or of faulty concepts of writing.

Frequently, writers have difficulty because they are not prepared to write. They simply have not done the work necessary to begin drafting their material with any kind of success. They have some vague ideas about what they want to say but have not worked out in any detail the necessary content and organization. Or they have worked out a sketchy outline when they need a full (perhaps a formal) outline. One of the real strengths of the design approach to writing is that it emphasizes and, in fact, insists on completing the tasks preparatory to actual writing. Writers who have carefully followed the earlier steps of the writing process should be well prepared to write their first drafts. They should have a good grasp of what they are going to say, why it has to be said, and to a degree, how it must be said. They can follow the designs they have developed with assurance and with clarity of purpose.

In addition to preparation, though, writing a draft, even more than selecting content and organizing it for presentation, requires extended periods of concentration. A few minutes at a time will not do. Interruptions, whether environmental or self-made, break writing periods into segments that are too short for effectiveness. Unfortunately, these interruptions that plague writers and discourage them from beginning at all are not so easily eliminated. Many elements in their environment create such interruptions, as do their own writing habits.

An argument in the office or just interesting social chatter can stop writing. The telephone, the events of the daily routine, and tangential assignments can make writing almost impossible. Consequently, many people who must do a considerable amount of writing often work for an hour or two after everyone else leaves the office or come to work an hour before starting time so that they can concentrate on their writing. This is one way in which to escape interruptions, to preserve a period of time for concentrating on the writing they must do. The real need, however, is for "quiet rooms," to which writers may go for uninterrupted periods of time, but until these become available, writers must devise their own ways to escape or to eliminate the external interruptions that plague them.

At the same time, writers should recognize how their own writing habits can cause troublesome interruptions. When they stop to revise sentences while writing a draft, they interrupt themselves. They stop the flow of their ideas by injecting the critical attitude. Writers who habitually work in this way are constantly interrupting themselves to do work that they can do better after the draft, or a large part of it, is complete.

Generally, writers who believe that they should turn out finished products at first writing often have trouble getting started. Expecting to do too much, they know that they will fail before they start. All that stands between them and effective writing is their own unrealistic expectations. They need to realize that the activity of writing includes both drafting and revising. They need to recognize that a different attitude is required for each and that constantly interrupting themselves while drafting their material interferes with their effectiveness. Learning to separate the act of writing from that of revising will do much to make writing not only easier but also more effective.

Faulty notions of introductions, including an overreliance on traditional beginnings, make it difficult for many writers to get started. Many writers can be trapped by their faith in such traditional introductions as "statement of the problem" or "background." Unfortunately, writing a meaningful statement of the problem is rather difficult if the report does not deal with a problem. Moreover, providing a complete statement of an assignment and report's background is also difficult if the writer is not aware of how or why the project arose. These traditional introductions, as valuable as they sometimes are, do not fit all reporting situations. Difficulty in getting started often arises from the effort to use them when they do not fit.

Traditional introductions are not the only ones that can trap the writer; the historical introduction is an even more frequent pitfall. Somehow, many writers have become attached to the idea that every piece of writing requires an elaborate historical introduction. Just how this misconception arises is not clear; it may stem, in part, from the overdependence on the chronological organization discussed in preceding chapters. But whatever its origin, it creates problems for writers, because they seldom are prepared to write a new Book of Genesis for each report. If historical information about a project is essential to a report, then the fact should become clear very near the beginning of the writing process, and writers should prepare to include it. In much of what scientists and engineers write, however, the historical introduction is unnecessary, and they should resist the temptation to attempt writing one.

Another faulty writing concept that makes getting started difficult for many writers is the belief that they must begin a draft on page one and continue to the end of the report. This notion, of course, confuses the order of presenting material to the reader with that of writing the draft. Actually, writers can draft the sections of a report in any order they wish, and they probably should start with sections for which they are best prepared to write and then move on to the others. Very often, they may find it easiest to write their introductory sections last. If they do draft the sections of a report out of the final order, however, writers must be sure to provide the necessary transitions from one section to the next.

NATURE OF THE WRITING ACTIVITY

It cannot be stressed enough that not only are the writing of a first draft and then revising it equally important aspects of the writing activity, but also they are two very different steps. Obviously, writing a draft is always necessary; otherwise, there would be no report. As for revision, on the other hand, too many writers believe that their first attempt at expressing their ideas should be good enough. The truth is that the longer, the more complex, the more important a piece of writing is, the more essential revision becomes. Sometimes, for short, routine pieces of writing, a first draft may serve as the final copy. But most of the time, achieving the accuracy and effectiveness necessary in technical writing requires revision—and frequently more than one revision.

More often than not writers do recognize that their first statements need revising, but many, if not most, of them attempt to draft their material and revise at the same time. As an almost inevitable result, the writing activity is slower, more tedious, less efficient, and more frustrating than it need be. The reason is that writing the draft and critically revising it are tasks having different objectives, involving different attitudes, and requiring different kinds of attention. Consequently, liberally mixing these two tasks creates all kinds of difficulty for the writer.

Writing is an act of shaping and giving expression to ideas in coherent language for the first time. It is, more or less, a creative act and requires a creative attitude. It requires the total concentration of the writer on a set of ideas, the initial statement of each, and the development of the ideas from one statement to the next. This movement from one idea to another, the momentum growing out of the association of ideas, is the key to the creative act of writing. The writer's problem is to keep the flow moving, rather than to interrupt it with any serious shift of attitude and attention. Stopping to examine critically and then revise a sentence while writing is exactly the kind of shift that can disrupt the flow of ideas, and making a habit of turning the flow on and off ignores the fact that creativity is not like a water faucet.

Revising is a matter of having second thoughts (and perhaps third and fourth thoughts) about the first statements of ideas. Its objective is to be sure that all the necessary ideas are on paper, that they do fit together in a meaningful way, and that the statement of each will clearly tell the reader what the writer had to say. Revising is criticism, and it requires a critical attitude. If writing demands a close association of writers and their material, revising requires that "distance" be established between writers and their statements. Before they can revise effectively, writers must let the "heat" of creativity dissipate; they must become as cool, detached, and objective about their drafts as they possibly can. At the same time, the momentum and flow of writing must be replaced by a slow examination and evaluation of the draft, from the reader's point of view rather than the writer's. Only by creating "distance" between themselves and the drafts they have written and by critically evaluating their statements can writers really see the inevitable differences between what they have said and what they should have said.

Much of the frustration caused by writing comes from the insistence on trying to mix the creative and the critical acts. Stopping in the middle of a paragraph to polish a sentence or in the middle of a sentence to search out the right word, writers often find that they have lost the idea they were going to include next. In addition, having stopped the flow of ideas into statements, they usually discover that they have lost their creative momentum and that this is not easily recoverable.

Writing the First Draft

When writers sit down with an outline and begin writing a draft, it is important that they concentrate on getting statements of their ideas down on paper. They should write as rapidly as they can and not stop to rework their statements. They need to build up "momentum," to capture the flow of associations in which one idea grows out of another and to maintain it as long as they can. Unfortunately, it will stop soon enough without their interrupting it.

Moreover, the flow will stop sooner than it would otherwise, unless they make some effort to hold onto the creative frame of mind and hold off as long as they can any serious intrusions of the critical attitude. There will be time enough for the latter. Indeed, separating the creative and critical work on a draft can result in getting more accomplished in less time than when the writer mixes them. This separation also can mean a better retention of ideas and an improvement in writing style. By writing rapidly and not stopping to revise every sentence as they go, writers may very well gain some ease and flexibility in their styles.

On occasion, critical questions occur to writers as they are working on a draft, and there is no reason they should completely ignore such questions. They can, and should, develop a system of notation by means of which they can quickly record on a draft the critical questions that arise. If they wonder whether a word is the one they really want to use, they can draw a circle around it. If they wonder whether more data, for example, or an illustration is needed at a certain point, they can make a brief note in the margin: data? example? illustration? If a paragraph or some sentences have possibly left a point unclear, they can draw a line in the margin alongside the questionable passage. Each time, however, a critical question comes to mind, they should do no more than make a record of it. They should not stop to resolve it. They should move on in order to maintain their momentum and come back later to do the necessary revision.

Revising

Before revising, writers should let their drafts "cool off," to create some "distance" between themselves and their newborn sentences and paragraphs. Time is an important element in this process, and letting a draft sit overnight before going to work on its revision is a helpful practice. There are others, as well. For many writers, having a draft typed causes their material to seem less familiar to them than it would be in the original version. Some writers also have their drafts typed on yellow second sheets. This practice works in two ways; it not only helps to create the "distance" mentioned but also persuades them to be less cautious in revising a typed manuscript. Knowing that the manuscript must go through the typewriter again, they are less reluctant to make changes in it than they would be if it had a more finished look. Something of the same effect can be gained if the manuscript comes back complete with strikeovers and other errors of the typist, who has been instructed not to be overly fussy in typing the draft. Typewritten manuscripts should, of course, be double spaced to leave some room for revisions, and if the finished reports turned out by an organization are normally single spaced, the double-spaced manuscript may encourage writers to revise freely. Incidentally, one of the best ways to get a cool, detached view of a draft, and helpful suggestions for revision, is to have a colleague review it. Writers should choose as their friendly critics, however, colleagues whose comments they are willing to heed.

Revising usually requires that writers go through their drafts more than once, looking for different kinds of weaknesses each time. They cannot expect to find all the necessary changes in one reading, unless the draft is very short. On the first reading they should check content and organization. Changes in these areas are often major ones and should be made before individual sentences are polished. During this first reading, writers should ask themselves a number of questions: Is there enough content of the right kind so that the report will achieve its purpose and serve its readers? Should any material be omitted? Is more explanation needed? Are relationships among the divisions and subdivisions clear? Is the order in which they come a functional one for the reader, or should it be changed?

After making the changes in content and organization which they believe necessary, writers should next read for accuracy, clarity, and effectiveness of statement. In this reading they should move slowly from sentence to sentence, examining and asking questions about each before moving on to the next. Are the data accurate? Are they interpreted for the reader? Will their significance be clear to the reader? Are the individual ideas presented so that their relationships are clear? Are the sentences well constructed? Are the words well chosen? Are any ambiguous? Can any be misunderstood? Can deadwood or roundabout phrases be eliminated? The elements of tone and style are discussed more fully later in this chapter. Here, it is sufficient to make clear that such matters are to be examined in this second reading.

After writers are satisfied with the statements of their ideas, they need to read the material again, this time checking for mechanics. This third reading is primarily one of correcting errors and inconsistencies in spelling, capitalization, and punctuation, perhaps; in the use of numerals and abbreviations; in the handling of equations; in tables and illustrations; and even in such things as pagination. Writers should also be on the lookout for inconsistencies or inaccuracies in references to figures or to tables. Matters of layout should also be checked to make certain that the heading structure truly represents the organizational patterns and relationships, that the layout of headings is appropriate and consistent, and that the headings themselves are clear and effective labels for the various sections.

Preparing Final Copy

When the report has been thoroughly modified and checked, the final copy that will serve as a basis for reproduction must be prepared. Consequently, the manuscript should be clearly readable so that the typists may easily and accurately make the reproduction or master copy. Writers also must give clear and complete instructions to the typists. The latter must know what kind of master copy they are to prepare —mimeograph, multilith, or a typed copy for a photo-printing process. They must know what form is to be employed and if there are any special problems, such as those frequently involved in the handling of equations, columns in tables, or captions for illustrations.

Writers should take special pains to make certain that all the instructions required are laid out clearly and that the typists understand them. If symbols are to be "inked in" after the report is typed, for example, the writer should be sure that a typist knows what these symbols are, where they go, and what amount of space must be left so that they can be added. If illustrations are being reproduced and will be inserted in the report after it is typed, the writer must tell the typist between what pages of text they will come or how much space will be required for each illustration that is to appear on a page of text. Writers, of course, must also give complete and accurate instructions concerning the reproduction of illustrations to those responsible for their preparation.

Producing a clean, clear, accurate master copy depends on a readable manuscript and clear and detailed instructions, and only the writer can see that these requirements are met. Whenever any question or doubt arises during the preparation of the final copy, the writer is the one who should be consulted.

The master copy must be proofread and all errors eliminated by the writer or by someone who is delegated this responsibility. If errors cannot be changed on the copy itself, pages must be retyped, for accuracy, neatness, and attractiveness contribute significantly to the effectiveness of any report. An attractively prepared report reflects favorably, not only on the writer and the organization, but also on the quality of the work being reported.

PROBLEMS OF TONE AND STYLE

Among the problems facing writers while they are writing and revising, those of tone and style are probably the most difficult. Stating a message that clearly serves its purpose and readers is no simple task, and often, in a hurry to meet a deadline, writers give too little time to expressing their ideas most effectively. Yet appropriate tone and effective style are as important to good technical writing as functional organization, and they should receive as much careful attention as any other element of a report.

The term *tone,* of course, refers to the attitudes of writers toward their subject matter and their readers, as these attitudes come through in what they say and how they say it. *Style,* on the other hand, refers to how writers say what they have to say, that is, to the kinds of words they choose and the kind of sentences and paragraphs they put together.

Both aspects of writing require attention, because each affects how much writers communicate to their readers. For example, if a scientist or engineer states a recommendation in unfamiliar words and fuzzy sentences, the readers are not likely to understand what the writer intended them to. On the other hand, a recommendation may be supported by statements that convey such an attitude as, "You should have had sense enough to take this action before." Then, annoyed by the tone, readers may not wish to act on a good idea that they understand very well. In either case, a vital part of the message did not get through.

Admittedly, most technical writing does not fall into either of these extremes. Nevertheless, it is effective only to the degree that its tone and style serve its purpose and help its readers to understanding and decisions.

In trying to solve problems of tone and style, writers should consider these problems before, during, and after the writing of their first drafts. Before they begin writing, they should make a determined effort to decide on a tone and style that will suit the purposes and readers of their reports. Do they, for example, wish to be friendly, informal, and simple in their expression? Or should they be coolly impersonal, using more formal and highly technical language? Whatever the decision, as they write their first drafts, they should consciously attempt to strike the tone that they have chosen and to select the words and construct the sentences that they believe will be most appropriate and effective.

Try as they will, however, they seldom achieve a completely satisfactory result in a first draft, no matter how skillful and experienced they are as writers. The concentration needed simply to get ideas down on paper leaves little opportunity for attending to many details of expression. On the other hand, once they have completed a draft, writers can give these details the necessary attention during revision. Thus they should allow enough time for careful revision of everything they write, from letters and memos to long reports. Revision should not be a hasty proofreading in which writers catch a few punctuation, spelling, and capitalization errors. Instead, it calls for painstaking attention to the many details that make the difference between poor tone and style and effective writing.

Tone

When they consider matters of tone, writers may face any number of choices, for, in theory, they can take up many attitudes toward their material or their readers. Yet in practice, the purposes of technical writing and the expectations of its readers limit the varieties of tone that writers may normally use. Certainly the occasions are few when they will want to be sentimental, whimsical, humorous, ironic, mocking, argumentative, satirical, hostile—the list of attitudes that most technical writing should not convey goes on and on. Indeed, it is so long that writers can forget that they still have some important choices to make in their day-to-day work. They should at least be aware of the effect they are likely to get with three kinds of tone: the impersonal, the negative, and the inflated.

Impersonal Tone

The first of these, the impersonal tone, results from using the impersonal style, which has become in this century the favorite of scientific and technical writers. Basically, the impersonal style represents an effort by writers to keep themselves and their readers out of their writing. They avoid using first and second person pronouns (*I, me, you, we,* and *us*) when it would be natural to put them in. At the same time, omitting these pronouns often causes writers to turn active verbs into passive ones. They may write, for example, something like the following passage:

> This alloy was selected for study because its properties and physical behavior in the formed and heat-treated condition are rather well understood.

Here, rather than use a direct and personal statement, "I selected this alloy," the writer chose to omit *I*. In doing so, however, he had to make two other changes: he had to turn a direct object *alloy* into the subject of his verb and then he had to substitute a passive verb *was selected* for an active verb *selected*. Similarly, instead of saying to his readers, "You understand its *properties* and physical *behavior*," the writer again turned things around and said, "Its *properties* and physical *behavior are understood*." In both cases, the writer expressed a detached attitude (an impersonal tone) toward his subject and his readers by omitting the personal references.

Although the impersonal style now seems absolutely necessary to many engineers and scientists, it has not always been *the* style for technical writing, and it is not without serious drawbacks. Its very impersonality of tone, the weak passive verbs, and the clumsy constructions that so often go along with those verbs can make technical writing more difficult and less interesting to read than it might be. As a result, some engineers and scientists are urging today that the impersonal style be abandoned in favor of a more direct and personal one, and the editors of many professional scientific and engineering journals are leading the effort to bring about this change. They argue that the impersonal style is not necessary for maintaining objectivity—that, in fact, scientists can be just as objective whether they write, "I selected this alloy," or, "This alloy was selected."

The argument is valid, of course, and the policy of abandoning the impersonal style for the personal one has much to recommend it. On the other hand, the impersonal style has a characteristic that frequently gets overlooked. What it can really do is to focus on *things*, and, very often, that focus, or emphasis, is exactly the effect the technical writer wants. In their professional work engineers and scientists are usually concerned with the nature, the manipulation, the causes, and the consequences of objects and processes. Moreover, when they get around to writing about what they have created or discovered, they can wish quite properly to emphasize the "things" that interest them and their fellow specialists. In their technical papers, articles, and reports, then, it may be appropriate for them to say, "This alloy was selected"; "The mixture was heated to 300°F"; "The shock tests were discontinued"; or "The following results were obtained."

Life would be simple if writers could choose once and for all between the personal and impersonal styles. Like many other either-or choices, however, this one oversimplifies a problem. If they are to be effective, technical writers must constantly be making decisions about using the impersonal style and tone.

For one thing, they should always distinguish between what the impersonal style requires and what it does not. Many writers, for example, become addicted to the passive verbs and fill up their reports with sentences like these:

> Capacitors sealed by molded plastics are affected by pressure changes.
>
> The design, development, and management of manned space systems will be facilitated through the use of mission simulation techniques.
>
> The reactor will be fabricated, assembled, and tested by ABC, Inc.

Yet such sentences do nothing to remove writers and readers from a direct relationship to each other or to the subject matter, or to emphasize "things." To maintain an impersonal style (and tone), the writers did not need the passive verbs *are affected, will be facilitated,* and *will be fabricated, assembled,* and *tested*. They might just as well have said:

Pressure changes affect capacitors sealed by molded plastics.

The use of mission simulation techniques will facilitate the design, development, and management of manned space systems.

ABC, Inc., will fabricate, assemble, and test the reactor.

Furthermore, by using active verbs, the writers could have gained some directness and vigor for their sentences. The gain may seem small in single examples, but if many writers would eliminate all the *unnecessary* passive verbs from their reports, the readability of their impersonal style would increase remarkably.

In addition to deciding what impersonality requires, writers should constantly be asking whether this style and tone are really appropriate to their purpose. Frequently today, engineers or scientists are not simply describing, explaining, or discussing the "things" of their professional concern. Instead, they can be requesting new personnel and equipment for a laboratory, defending a budget for a piece of research, outlining a policy or a work schedule, or answering a question about how one of their projects will affect a taxpayer. In short, they often can be writing letters and memos, even reports, on any number of subjects that may seem incidental to their work but are necessary to getting it done. On most of these occasions, the impersonal style is not appropriate, for the focus should not be so much on "things" as on the *human affairs* that directly concern writer and reader. Nevertheless, many specialists have grown accustomed to writing impersonally about their work, and they continue doing so when circumstances require a different attitude. The unhappy results show up particularly in the correspondence of technically oriented organizations. All too common are letters like the following paragraphs with which one man opened a request for the help of some consultants:

Although United Foundry has been in operation for only a few years, its present facilities will soon be outgrown. It is therefore planned to build an additional plant at a site not yet determined.

Since this firm has had no experience in making plant location studies, Engineering Consultants are being asked for professional advice, in order that the most suitable location for the new plant might be chosen.

Aside from having some vague and especially clumsy passages, these paragraphs give little evidence that one group of human beings is asking another group of human beings for help. By omitting all personal references, the writer insisted that he, his reader, and the other human individuals that they represented were of no consequence to his message. More that that, he treated a human decision (where to locate a new plant) and the human relationship that he wanted to establish as if they were "things." These obvious attitudes are not appropriate to his purpose; they actually work against it. The letter's tone, in fact, scarcely suggests anything resembling human concern, friendliness, warmth, or even courtesy.

Very often today, large organizations in government and industry get criticized for being cold, faceless monsters having no regard for human individuals. Apparently to offset this criticism, many companies and agencies spend large budgets on personnel relations and public relations programs designed to prove that they are just big, happy families of warm, friendly people. Yet in these same organizations, engineers and scientists (indeed, every kind of specialist) go on writing memos to one another as if they really were cogs in a machine. At the same time, the impersonal letters that they write to suppliers or customers or to the general public demonstrate again and again that the critics of their organizations must be right. What is more serious, if a few writers try to make their letters and memos sound human, they are likely to have a manager whose constant reminder is that "we just don't write that way around here."

So long as an organization's "friendly people" (and they usually are) continue to write impersonally at the wrong times, the efforts of the public relations experts seem useless. Many organizations might do well to care less about creating a good "public image" and to care more about encouraging their employees to adopt a direct, personal style in correspondence that is truly representative of real human concern.

If the man who wanted help in locating a new plant had tried a more direct and personal approach, he might have opened his letter like this:

Although our firm has been in operation for only a few years, we will soon outgrow our present facilities. Consequently, we plan to build an additional plant at another location. Because we have not had experience in making plant location studies, we are seeking your professional advice, in order that we might choose the most suitable location for our new plant.

This improved version may win no ribbons for being great prose, but it has more than one advantage over the original. Using the first person *we* and the adjectives *our* and *your* and getting rid of several passive verbs have helped to make the revised letter less clumsy and more clear than the first one. In addition, the *we, our,* and *your,* no matter how indefinite a group each represents, include writer and reader in the message. Introducing this personal element gives some sense that one living person is asking another living person to share a human problem.

On many occasions, writers can and should use *I* in correspondence. An engineer, for example, may be solely responsible for observing how well a factory manages its quality control program. When he gets around to writing his boss a memo about what he has learned, he has no reason (except the boss, of course) to hide behind the impersonal style or to use some clumsy device like *the writer, the undersigned,* or the editorial *we*. He might just as well accept his responsibility and say, "I found these to be the facts," and "These are my conclusions."

In other situations the use of first person can seem more open to question. The scientist interpreting for a taxpayer a government regulation on pesticides might argue that she should write her letter using the indefinite *we,* because, after all, she merely represents her agency. On the other hand, the indefinite *we* can include everyone in the agency from the scientist to the director, and, with careless use, this pronoun can become as false sounding as the impersonal style. Taxpayers, like many other readers, have a personal interest in their problems; they appreciate attention that sounds personal; and, with reason, they can be annoyed almost as much by "We believe" as by "The agency believes" or "It is believed." Ordinarily, those who sign letters have some kind of authority to speak for their organization. When they do so, they should not lose themselves in a *we* so that they can reduce their personal involvement or simply avoid using *I*.

Contrary to a common prejudice, the pronoun *I* in business correspondence is not necessarily a mark of egotism or of overly great ambition. Almost all persons use this pronoun quite naturally when they speak directly to other persons, and in a letter it helps to suggest that the writer is paying some close, personal attention to the reader and the subject at hand. As for use of the personal style generally, it need not lead to overfamiliarity, false heartiness, or aggressive friendliness—the kinds of tone one so often finds in direct-mail advertising. These are abuses of the personal style in correspondence; writers do not make such mistakes just because they decide to use *I, we,* or *you*. How they use these pronouns, what other words they choose to go with them, and how they put everything together into sentences and paragraphs make the difference between success and failure in a letter's tone.

Engineers or scientists, then, should not restrict themselves to the impersonal style; its tone is clearly wrong for much that they write. At the same time, the important question may not be whether they should use a personal style, but how "personal" they should get. In the technical report the impersonal style can be appropriate, but in ordinary memos and business letters it seldom is. Thus, for these kinds of writing the choice between the personal and impersonal styles should be reasonably easy, provided, of course, the writer has freedom to choose. Between technical reports and correspondence, however, lies a grey area of other reports, directives, instructions, procedures, and the like, for which one style or the other may be appropriate. In this area particularly, technical writers may find the following guidelines helpful:

1. The impersonal style and tone are useful when the purposes of writers require that they emphasize the physical objects, events, and processes that they have investigated or produced.
2. Use of the impersonal style should not tempt writers into choosing unnecessary passive verbs.
3. The personal style provides the human tone that is so important when writers are taking up human affairs, and particularly when writers and readers are directly concerned in those affairs.
4. In the personal style, writers should use *I* and *you* when it makes good sense to do so.
5. Writers should remain consistently with the impersonal or personal style unless they have a very good reason for mixing them. Careless mixtures of the personal and impersonal, as in "your recommendations have been received," create contradictions in both style and tone.

Negative Tone

The second tone with which engineers and scientists can have serious problems is the negative one. Very often they convey to readers a negative or pessimistic attitude toward their subject matter when they do not intend to do so. Almost as often, they can have trouble with tone when they must say "no" to a reader.

Generally, an unconsciously negative attitude toward their subjects seems to be more common in the writing of engineers than in that of scientists. Perhaps the practice of engineering encourages pessimism, or perhaps it teaches great caution about what one writes down for others to see.

The tendency to find defects and failures more noteworthy than virtues and successes is probably a useful characteristic for engineers. After all, a large part of their jobs can be to ensure that complex physical structures serve other human beings reliably—that bridges stand, engines run, and airplanes fly. Consequently, the final detail preventing success can have more importance to engineers than all they have done before. They have not really done their jobs until "the thing works" as well as it must. Possibly, too, engineers learn more from their failures than they do from their successes. If they have tried to build a good bridge and it collapses, they are more likely to learn something new about designing and constructing bridges than if it had remained standing. Nevertheless, no matter how useful their taste for trouble may be in their professional work, they should not permit it to distort what they write.

A mistakenly negative tone can result from the cumulative effect of an entire report, or it can sound loud and clear in a single sentence. A man who had directed qualification tests of a battery that his company wished to use in a complex system for one of its customers began the opening summary of his test report with the following sentence:

> Two of the 32 specimens tested failed to meet all of the test requirements.

Now, although the sentence does not say that the tests proved the battery to be unacceptable, it strongly suggests that the keynote of the report will be failure. The writer's statement was true enough, but he chose to emphasize some very minor defects that his tests had uncovered in 2 of 32 sample batteries. Unfortunately, the resulting negative tone started the reader off with the wrong attitude, because the report was intended to inform the customer that the battery was acceptable. Indeed, by the standards the customer had laid down, it had passed the qualification tests exceptionally well.

To make matters worse, the writer kept his reader in the wrong frame of mind throughout his report. In the remainder of his summary, he carefully described each defect and worked very hard to explain each one away. Only near the close of his summary did he state, almost as an afterthought, that the battery had qualified. As for the body of the report, it was simply an enlarged version of the summary. Not too surprisingly, after an engineer representing the customer had looked over this report, he returned it, and he added the advice, "If you want my organization to approve the use of this battery, tell us something good about it. Don't just tell us everything that's wrong with it."

The writer, naturally, had to overhaul his test report completely. In doing so, he revised the first sentence of his summary to read:

> Thirty of the 32 specimens tested met all of the test requirements.

This statement takes in exactly the same facts that the original sentence included, but it expresses a positive attitude toward them and thus represents them more fairly. Once he had started with a positive tone, the writer tried to maintain it until the close of his report. He did not, incidentally, neglect to discuss the defects that he had found, nor did he attempt to slide over them as quickly as he possibly could. However, he did give the defects less attention than he had given them originally, and he was less defensive about them. At the same time, he had more to say about how well 30 of the sample batteries had performed. With his revision completed, the writer had a report that included all of the data and conclusions in the first one but was completely different because of its positive tone. When the writer submitted his second report, the customer accepted both it and the battery.

Occasionally, engineers feel that any advice to shift from a negative to a positive tone is a suggestion that they misrepresent what they have done or learned. They should realize, though, that adopting a positive tone does not necessarily mean that they must ignore failures or deficiencies. Certainly, if these are so serious that they ought to be the focal point of a report, then using a positive tone in an attempt to deceive the reader is being dishonest.

However, the point is that although it may not be dishonest, a negative tone misrepresents the facts whenever it contradicts the engineer's best judgment about them. If defects and failures are but some of the facts a report should contain, the writer should give them no more attention and emphasis than they deserve. They should not be stressd at the expense of the positive evidence from which the writer concludes that a project is advancing or a battery will do the job required of it. To put the matter another way, a positive message should have, primarily, a positive tone.

The mistakenly negative attitude toward subject matter may prove most troublesome to engineers, but scientists and engineers generally have their problems when, in correspondence, they must react negatively to a reader. Far too often a letter or memo's "no" to a request, for example, can be much stronger than necessary. The main reason for this difficulty apparently is that writers can fail to see much beyond what concerns them directly; consequently, they can be indifferent to the reader's needs and can ignore any claims that the reader might have to thoughful consideration.

A citizen wrote to a government agency about some information that he needed. He had, it seems, the opportunity to buy some land near a small town, and he wanted to learn if the water backed up by one of the agency's proposed dams would cover this property. So he asked if the agency could send him a map showing the information he needed or if it could tell him where he might get one. The hydrologist who replied for the agency responded truthfully that the agency had no such maps and they were not available anywhere. Then he added three paragraphs explaining, often in rather technical terms, why it was impractical to prepare this kind of map. When the citizen received the hydrologist's letter, he became angry enough that he fired off a demand to know why his legitimate request for information had gotten a double-talk brush-off.

Although the tone of his second letter did little to help communication, the citizen did have justice on his side. First of all, the hydrologist had said "no," and then he had amplified the sound of his "no" with every reason that he gave to explain why the citizen should not expect to get a map. Adding to this amplification was the fact that the citizen could not understand all that the hydrologist had said, and he did not see how the hydrologist's problems had much to do with his own. Second, and even more important, the hydrologist had answered only the specific question about the map. He failed to recognize that the map was secondary, a means to an end. What the citizen really wanted to know was whether some land would finally be above or below a waterline. In effect, the hydrologist's letter was triply negative: Not only did it say "no" more loudly than it had to, it suggested clearly that the reader's problems did not concern the writer at all. He might have done better by his reader, himself, and his agency if he had written:

> The kind of map you want is not available anywhere because it is too difficult and expensive to make. However, if you will let me know the exact location of the land you want to buy, I probably can tell you whether or not it will be under water.

The overly negative tone of the hydrologist came principally from what he did and did not say. On other occasions, writers can be careless in phrasing a "no," and their tones can take on a nasty edge that they might not be entirely aware of. For example, an industrial recruiter wrote to the director of an engineering college placement office about arrangements for his coming visit to the campus. In the course of his letter, the recruiter asked for an increased use of the office facilities. With mounting anger, he read the director's reply, which began:

> I am in receipt of your letter of February 1 and only wish to correct one thing and this is—we do not have four representatives from any company at this office simultaneously unless extraordinay circumstances warrant it. We schedule 10 companies at one time and our physical facilities and personnel do not allow for any one company to overburden them to the exclusion of the other organizations we serve.

The first point to note about this paragraph is the especially clear example it provides of the close relationship between style and tone. When writers insist that they have only one thing to tell a reader, the chances are very good that they intend to hit the reader between the eyes—and this, of course, is just what the director did. The second point is that the director probably realized he was saying "no" emphatically, but he could very well have missed his self-righteous overtones of outraged virtue and schoolmasterish rebuke. Finally, what the director said and implied may have been exactly what he needed to say. Perhaps the recruiter did ask for more than the director's policy could allow if the office were to be fair to other employers. However, the manner in which the director phrased his refusal produced an insulting tone that lost him the goodwill of an important company representative. In cases like this one, it can be difficult to pin down just where the tone goes wrong. Obviously, much of the effect comes from the choice of such words and phrases as *correct, and this is, do not, extraordinary circumstances warrant, any one company, overburden,* and *exclusion.* Nevertheless, it probably would take a complete revision to get a satisfactory paragraph.

Incidentally, the director continued in several more paragraphs to express much the same attitude he had in the first. But at the close of his letter, he produced a complete contrast in tones by falling back on some stereotypes that are common to correspondence. He began his final paragraph with the formula: "It is a pleasure to serve your fine company"; here, though, he could not resist turning the knife one more quarter turn: "along with the hundreds of others we serve annually"; and then he closed with the formula: "I remain/Cordially yours." If the recruiter had not been angry by this time, he might have laughed at this quick trick of the writing hand.

Sometimes, perhaps, a letter or memo must say "no" to its reader and do so emphatically, even sharply. Certain readers can be unreasonably slow to understand. However, writers should always be sure that the force and sting of their "no" really do fit the purpose of what they are writing. Misjudging these qualities of tone is all too easy when scribbling down or dictating a first draft. In fact, writers can remain entirely unaware of such qualities while they attempt to get their ideas on paper—just as they can remain unaware that they are being negative about their subject matter. Thus when writers must say "no" or discuss failures and deficiencies, they should take particualr care in checking their first drafts. They should try to "hear," as the reader will, the tone of what has been said, and they should revise with care whenever the tone becomes needlessly or mistakenly negative.

Inflated Tone

A third tone to which writing scientists or enginners should pay careful attention is the inflated one associated with a style often called jargon. It is only fair, of course, to insist that engineers and scientists hold no copyrights on jargon and inflated tone. These appear everywhere in the writing of a world full of experts. Interestingly enough, when experts talk, they can, and very often do, speak in relatively direct, simple, down-to-earth language. But put pens or pencils in their hands, no matter how modest their special knowledge, and the experts' language can suffer from a ballooning effect that threatens to lift their messages out of all contact with the real world.

A classic example of inflated tone appeared during World War II in an order from the Office of Civilian Defense. Concerned with blacking out federal buildings during air raids, the order stated at one point:

> Such preparations shall be made as will completely obscure all Federal buildings and non-Federal buildings occupied by the Federal Government during an air raid for any period of time from visibility by reason of internal or external illumination. Such obscuration may be obtained either by blackout construction or by termination of the illumination.

As the story goes, this order reached the desk of President Franklin Roosevelt, and he recommended some deflation:

> Tell them that in buildings where they have to keep the work going, to put something across the window. In buildings where they can afford to let the work stop for a while, turn out the lights.[1]

What compelled the person who drafted that OCD order to write this way? Or, what generally tempts the expert when writing from the authority of special knowledge to use such language? There are, probably, no simple answers to such questions. However, most readers exposed to prose of this kind arrive at some unflattering conclusions about those who write it. They quickly discover how clumsy and fuzzy such prose frequently is, but they also can "hear" how stuffy it "sounds," at its best, or how riduculously pretentious, at its worst. As a result, readers often conclude that writers are attempting a "snow job," are showing off how much they know, or are naively trying to make their messages sound important, perhaps more important than the messages actually are. Needless to say, an inflated tone seldom impresses any but the most uncritical readers. As for the others, it amuses them sometimes. More often it irritates them into ignoring what a writer has to say, or if they cannot avoid reading it, the inflated tone quickly invites their mockery and skepticism.

Style

Many writers do not associate a serious concern for style with scientific and technical writing. When they think of style, they think of imaginative writing, of novels and poetry. Yet a satisfactory style is fundamental to the effective presentation of ideas in scientific and technical writing, and certainly the style of much of that writing can be greatly improved.

Actually, the style of a report results from the cumulative effect of a great many choices—choices of words and phrases, of patterns for combining these words and phrases into sentences, and of the larger patterns for linking sentences into paragraphs. In writing a report, the number of such choices is very large; consequently, identifying and removing the causes of an unwanted stylistic quality are not always simple matters to be taken care of with little effort. They usually require careful attention to the many details of language that make the difference between styles that are weak or strong, cold or warm, easy or difficult to read, roundabout or direct, graceful or awkward, personal or impersonal, concerned or indifferent.

[1]Quoted from Porter G. Perrin, *Writer's Guide and Index to English*, Scott, Foresman and Company, New York, © 1942, p. 222.

An important point to remember is that "correct" language alone does not ensure that a style will be satisfactory. Of course, use of the grammatical conventions accepted by educated persons represents the foundation on which to build an effective style. Few, if any, report writers want their readers paying more attention to their grammatical "errors" than to what their reports have to say. On the other hand, too much scientific and technical writing can be grammatically "correct," yet also be dull and ponderous, even impossible to read. A concern for style, then, goes beyond the "rules" of grammar to a consideration of the many other aspects of using language that are important to stating ideas effectively.

Writers probably should give the stylistic element of their writing the most attention during revision, for in the "heat of composition" they must attend to getting their ideas down on paper. Only during the "cool" period of revision can they give full attention to the details of language that can produce an effective style. In revising their drafts for stylistic improvements, writers may find it helpful to work on various levels of their prose at different times; that is, they might give attention first to the effectiveness of their paragraphs, look next at their sentences, and finally consider their choices of words. How much writers divide their labor on style in this way depends, naturally, on how long the drafts are and how much time they have for revision.

Paragraphs

At the paragraph level writers need to make certain that they have presented their material in comprehensible blocks of content—that each paragraph has unity (it adequately develops a central idea or concept) and that each has coherence (all its statements hold together in close and meaningful relationships). In technical writing the unity and coherence of paragraphs can be weak because writers rely too heavily on the short, simple sentence or they fail to use transitional words and phrases. Of course, the overuse of simple sentences (listing ideas) and a lack of transitions may not be the only flaws to be looked for in the paragraphs of a draft. Nevertheless, they are among the commonest weaknesses in technical writing, and, especially during revision, writers should especially be on the lookout for them.

Listing Ideas. Much too often in scientific and technical writing, undue respect is paid to the value of the short and simple sentence. Presumably, the theory is that the shorter and less complicated each sentence, the better the communication. Unfortunately, writers who insist on isolating each one of their ideas in a short, simple sentence do little more than present their readers with a uniform *list* of ideas. When all the major and minor ideas of a paragraph appear in the same kind of brief statement, there is nothing to distinguish among them. In effect, such writing has neither foreground nor background; it lacks perspective because each idea gets the same emphasis. And, in the "flat" paragraph, readers can have very great difficulty in discovering the central concept that it may be attempting to develop.

Looking closely at the listing of ideas from a somewhat different angle reveals that it tends to suppress the important relationships among a group of ideas that give them their full meaning. If, for example, a writer sees a causal relationship between two events, it makes little sense to say, "The gear broke. The machine stopped." Readers would find it much more helpful if the passage read, "The gear broke. Therefore the machine stopped." Here the use of the transitional word *therefore* does relate the ideas in the two simple sentences (and the next section discusses the value of such words). However, the causal relationship can be expressed most directly and firmly in a single complex sentence that subordinates cause to effect: "Because the gear broke, the machine stopped." This sentence, of course, is more than twice as long as each of the first two simple sentences, and according to some theory, it should be more than twice as difficult to read and understand than they are. Yet a reader's experience indicates that the reverse is true. Using the kinds of sentences that clearly express the relationships among ideas will increase the coherence of a writer's paragraphs and hence their readability and understandability.

Overuse of the short, simple sentence produces a style that aptly has been called "primer style," a term derived from the kinds of sentences to be found in first-grade reading books: "Jane saw the dog. The dog's name was Spot. Jane called Spot. Spot ran to Jane." Why passages somewhat like this one are frequent in technical writing today is rather diffucult to understand, because the thought underlying such writing has much more complex and subtle relationships than anyone would try to convey in a reading primer. Perhaps the following paragraph can demonstrate the weakness of primer style, even though its underlying body of thought is not overly complex. In the paragraph, the writer wished to develop the central concept that a contractor is putting in as much work on a construction site as before, but that dry weather is causing delays in some parts of the job. What he wrote took shape as four simple sentences and a complex one:

> The contractor is still operating in two shifts. The debris from the clearing operation is now being piled. No burning at all is being permitted. The continued dry weather has produced extreme fire hazard conditions. No power equipment is permitted to operate when the relative humidity is below 20%.

Immediately after finishing this paragraph, readers may easily conclude that it has no unifying concept at all. Because almost every idea stands alone in basically the same sentence structure, most of the ideas seem exactly alike in importance. Paradoxically enough, however, this very sameness causes the ideas to seem different enough from each other so that the readers can wonder if each new sentence is a new beginning for the paragraph.

If they take the time to go through this paragraph again, the readers may come to the uneasy realization that all its ideas are related in some way. Indeed, they may guess that the idea stated in the first sentence stands in some sort of contrast with the following ideas and that the latter probably fit into a pattern of cause and effect. Nevertheless, the readers can only guess at these relationships because the writer's simplistic style eliminated them.

To overcome the effect of listing ideas, writers should adjust the length and basic structure of their sentences so that they include as many as they can of the relationships they see among their ideas. Very simply, this kind of revision usually requires combining short, simple sentences into longer ones of a different type. Two simple sentences may become a compound sentence if they present two closely related but equally important ideas. Two or more simple sentences may be combined into a complex sentence whenever some ideas should be subordinated to others. On occasion, the compound-complex sentence provides the best means of expressing an especially complicated set of relationships. Sometimes also, combining sentences effectively may require that the order in which ideas appear be changed, or it may require that some ideas be stated more fully than they were in the draft. In any event, the revision of primer style finally should result in a variety of sentences, ranging from simple to compound-complex (perhaps) and from short to long (but not too long).

However they proceed in revising passages of primer style, writers should not go to an extreme. The objective of combining sentences is not to eliminate all short, simple sentences from a draft; they are especially useful as a change of pace in the flow of sentences and for emphasizing some ideas. At the same time, criticizing the view that the simplest sentences always communicate best does not imply that overly long and involved sentences ever communicate well. They do not. For this reason, writers need also to keep in mind that there are limits to how many sentences they should reorganize into one.

These observations, incidentally, apply to the paragraph introduced above as an example of primer style. It is possible to relate the ideas of this paragraph reasonably well by combining its five sentences into one. Doing so, though, would not be particularly desirable; the result probably would be pushing the limits of what readers should be expected to grasp in a single statement. A better approach to improving the paragraph might be to combine some of its sentences and then to rely on the transitional elements *however* and *in addition* for linking other ideas—as in the following revision:

> The contractor is still operating in two shifts. However, the debris from the clearing operation is now being piled because, under the extreme fire hazard conditions produced by the continued dry weather, no burning at all is being permitted. In addition, no power equipment is permitted to operate whenever the relative humidity is below 20%.

This revised paragraph is not perfect, of course, but it will communicate more than the original.

Transitions. As primer style demonstrates very well, disjointed writing does not present ideas effectively. This style, however, is only a special and extreme case of the failure to relate ideas. Such failure occurs wherever readers can honestly wonder why one sentence follows another, and first drafts are likely to have any number of these puzzling gaps. Thus during revision writers should always be looking for those points at which they must bridge a gap between their sentences. Doing so may take only a word or a phase; at other times, it may require a clause or even a sentence. In any case, writers will want to choose their transitions carefully, for good transitions represent one of the most important differences between disjointed paragraphs and coherent ones.

Just where writers need to add transitions is a question that cannot be answered specifically until they actually sit down to revise a draft. Yet some general discussion of how they might go about seeking an answer may prove helpful. As a working principle, writers should accept that a draft's entire sequence of sentences is a chain of thought in which each statement should follow unquestionably from the statement before. This view implies, of course, that each new sentence should have an obvious link with its predecessor. Moreover, the linkage should be made explicit, for it does not matter if writers understand the relationships between their sentences; the important question is whether readers can see these relationships. All of this is to say finally that a need exists for some kind of transition, however slight or subtle, between every paragraph and between every sentence within each paragraph.[2] Consequently, the goal in revision is to make sure that the need has been satisfied in every case.

To satisfy the need effectively, writers can use a number of different techniques. Some of the more common are listed below, with examples drawn from the paragraph immediately preceding this one:

1. Rephrasing the main idea of the paragraph that has gone before (''Just where writers will need to add transitions'') or the idea of its final sentence.
2. Repeating key words of the preceding sentence or sentences in the same or in a different form (''writers,'' ''transitions,'' ''answer,'' ''sentence,'' ''linkage,'' ''need'').
3. Using pronouns to refer to key words in preceding sentences (as ''they'' of the second sentence refers to ''writers'' of the first).
4. Using transitional words or phrases that refer directly to an idea (''This view'') or a set of ideas (''All of this'') that has gone before.
5. Using transitional words or phrases that express relationships between ideas (''Yet,'' ''Moreover,'' ''Consequently''). The language has any number of such expressions; other examples include ''however,'' ''nevertheless,'' ''therefore,'' ''at the same time,'' ''on the other hand,'' ''in the first place,'' and so on.

[2]Many books on writing draw a distinction between *connecting* sentences within a paragraph and supplying *transitions* from paragraph to paragraph. In both tasks, however, the principles and practices are largely the same. Thus, for the sake of convenience in this brief discussion, the term *transitions* refers to the means for linking both sentences and paragraphs.

Writers who wish to improve the transitional quality of their writing, to link their sentences and ideas together more effectively, should take time to examine writing they find readable and coherent to determine how others relate ideas and signify relationships for the reader. A few minutes examining the linkage in effective writing will reveal many transitional techniques and linkage devices.

Sentences

At some point in revision, writers should look carefully at all the sentences that make up their drafts. They should examine each one to ensure that it states its idea clearly, concisely, and straightforwardly. To achieve this obective requires, of course, that they consider a variety of ways in which the quality of sentences can be improved.

Certainly, writers should want to be sure that the structure of their sentences is grammatically ''correct.'' Faulty grammar, as noted earlier, calls unnecessary attention to itself; moreover, faulty sentence structure, by upsetting the normal expectations of readers, can obscure meaning and confuse them. Usually, their confusion is only momentary. Readers can pause and straighten out the faulty constructions that writers should have corrected during revision. On the other hand, a steady accumulation of faulty structures produces a style that readers find increasingly irksome. Meaning seems continually to be slipping away from them, and the repeated efforts to grasp it make reading a heavy chore to be avoided at every possible opportunity.

Writers should also want to see that their sentences are as concise as the clear statement of their ideas will permit. Curiously enough, despite the stress on fog indexes, short sentences, and brevity of presentation in industry and government, wordiness is probably as common in reports today as it ever was. Perhaps, however, this fact should not be too surprising, for the secret of writing concisely is not to be found in the magic length of short sentences, short paragraphs, or short reports. Rather, it is to be found in eliminating from a draft those words and phrases that are truly unnecessary to the meaning of each sentence. Writers can achieve conciseness by sharpening their eyes for the specific ways in which they can waste words —deadwood, roundabout phrases, and indirect sentence openings.

Finally, writers should seek to reduce the density of sentences in which they have listed, rather than related, details. Just as they can overuse the short, simple sentence and hence list ideas, so writers can overuse the prepositional phrase and pile detail on detail, until the relationships among the parts of an idea become difficult to understand.

Needless to say, it is one thing to recommend that writers should revise faulty sentence structures into sound ones, wordiness into conciseness, and listed details into related ones; it is quite another to carry out these recommendations. Thus the following discussion of specific examples attempts to illustrate what writers can do during revision to improve the quality of their sentences.

Faulty Structure. Unfortunately, at its very beginning, this discussion faces a major difficulty. To identify the numerous ways in which the writing of English sentences can go wrong and to suggest the means for correcting each fault normally requires a major section in an English handbook. To present such coverage here, however, is neither possible nor, for that matter, very practical. Indeed, writers who find that their command of sentence structure is shaky would do well to obtain one of the many handbooks available and consult it regularly while revising their drafts. In any event, what follows is limited to discussing three kinds of faulty structure that are particularly common in scientific and technical writing—the vague reference of pronouns, the dangling modifier, and the misplaced modifier.

To be clear in their meaning, the personal pronouns *he, she, it,* and *they;* the demonstrative pronouns *this, that,* and *these;* and such relative pronouns as *which* must refer to specific nouns which usually precede them in the same sentence or in the sentence that has gone just before. When readers cannot find a noun to which a pronoun might sensibly refer or they are confused about which noun is being referred to, their efforts to make sense of what they are reading are baffled, if only momentarily. Of course, in writing a draft writers may very well leave unclear the reference of some of their pronouns, and unless they remedy this situation during revision, their readers can confront sentences like these:

1. Without water this land would be virtually desert, *which* would result in heavy losses.
2. The assigned values reflect the absolute maximum that could be considered, and if I were doing *it,* I think I would be from 25¢ to 50¢ lower.

In both cases, readers have very much the same problem, for the *which* in the first example and the *it* in the second have no possible antecedents at all. Irritated readers with time on their hands can demonstrate this fact by substituting all the available nouns in each sentence for the pronoun in question and obtaining no sensible results. But the real difficulty is that every reader must guess at the meaning of *which* and *it,* and consequently, at the meaning of each sentence. No doubt, readers can do so with a good probability of success. They may very well conclude that the writer of the first sentence really meant to say something like this, "Without water this land would (become) virtually desert, (a change) which would result in heavy losses." And they may decide that the second writer meant to say, ". . . and, if I were (assigning the values), I think I would be from 25¢ to 50¢ lower." The important point for writers to remember, however, is that their readers' guesses are only guesses (no matter how good) and that such guesses require the labor of mental revision they could have made unnecessary with their own careful revision.

The problem of vague reference needs particular attention from writers who liberally sprinkle the pronoun *this* throughout their writing. Much too often in writing, the meaning of *this* is vague at best and uncomprehensible at worst. Unfortunately, although the word seems a convenient way in which to point at much that has gone before, this very convenience requires that it be used with care. For *this,* as for every other pronoun they use, writers should make certain that the antecedent will be unmistakably clear to their readers.

The dangling modifier, so called because the modifier—word, phrase, or clause—has no clear connection with the sentence structure, will also cause difficulties for readers. The difficulties may seldom be major; that is, readers probably will not completely misunderstand what a writer has said. Nevertheless, danglers do create puzzles of meaning, cause a reader to pause for some mental revision, and, if they occur frequently, produce an undesirable fuzziness of style.

Interestingly enough, the dangling modifier often comes at the beginning of a sentence, and it leads the reader to believe that the sentence is moving in one direction until, in fact, the sentence moves off in another. This kind of shift is illustrated in the following example:

In reviewing the project records, there was no record showing that the foundation compaction was obtained.

Here the italicized introductory phrase causes the reader to expect that the subject of the main clause will designate a person or group of persons who did the reviewing. Such an expectation is usual among educated readers, at least, because in standard English usage the implicit agent of an action like reviewing normally does prove to be the subject of the main clause. However, after the delay of an indirect opening ("there was"), the subject of the main clause in this case turns out to be *record,* and readers have the choice of accepting or rejecting the notion that a nonexistent record reviewed some records. Readers may find the notion amusing, possibly, but very likely will revise the sentence to read, "In reviewing the project records, (I, we, he, she, they, the committee, or whoever found) no record showing that the foundation compaction was obtained."

Perhaps no modifier gets dangled so frequently as the participle, a verb form (*storing, stored, having been stored,* for example) that serves to modify nouns and pronouns. Of participles, readers expect (1) that each one will clearly and sensibly modify a noun or pronoun in its sentence and (2) that a participle coming before the main clause will modify the subject of that clause. Unfortunately, writers who are careless in revision can upset the expectations of their readers with sentences like this one:

Stored at 42° Fahrenheit, bacterial counts increased greatly.

Now the only possible way in which the past participle *stored* can fit into this sentence is to modify the noun *counts*. In other words, the counts were stored. But is that what the writer meant to say? Readers, wiser than the writer in this case, are not likely to accept this piece of science fiction; puzzled for a moment or two, they are much more likely to conclude that on some samples stored at 42° Fahrenheit, bacterial counts increased greatly.

As the discussion of dangling modifiers has suggested, their position in a sentence can affect their meaning, and this matter of position is important to modifiers generally. In the English sentence modifiers are usually placed close to the words they modify. If they are not, if the distance between a modifier and the word it modifies becomes too great, the result is a misplaced modifier and confusion again for the reader. The two following examples illustrate the problem:

1. Short distances between the highway and the itnerchange would permit continuous illumination and avoid the succession of light and dark areas that would otherwise occur *as recommended in the design report*.
2. We will reimburse you after reporting to work *for a first-class family airline fare* between Los Angeles and the airport nearest your home.

In the first example readers are surprised to learn "that the succession of light and dark areas" to be avoided would "occur as recommended in the design report." They will, of course, immediately question the sense of that connection and recognize that the phrase *as recommended in the design report* must modify some other word in the sentence than the verb *occur*. At this point, however, revising the sentence encounters a slight difficulty, because readers cannot tell whether the phrase should modify the verb *permit,* the verb *avoid,* or (somehow) both. The differences of meaning represented by these choices may be small. Nevertheless, the sentence is one that the writer probably should have started all over again while revising the draft.

If he has a sense of humor, the man who received the letter including the second example might still be chuckling over what that sentence says. Placed where it is, the phrase *for a first-class family airline fare* apparently modifies *work,* and, hence, the sentence seems to contain a ludicrous announcement about what the man's pay envelope will contain. No doubt, the reader quickly saw that the phrase was intended to modify the verb *reimburse,* but that it had been separated from this verb by the phrase *after reporting to work*. The letter writer could have avoided some chuckles at his expense by taking the phrase separating the two parts of the sentence that need to be brought closer together and placing it at the beginning of the sentence. Being careful not to dangle *after reporting,* the writer could have revised the sentence to read:

After you report to work, we will reimburse you for a first-class family airline fare between Los Angeles and the airport nearest your home.

The misplaced modifier, as the vague pronoun and the dangling modifier, can occur with unhappy frequency in the sentences of a draft. The writer's task is to spot these faults and then to revise or rewrite sentences so that readers need not pause to do some revising of their own.

Wordiness. Being concise, rather than wordy, requires that writers distinguish betwen (1) those words and phrases that truly convey meaning to the reader and (2) those expressions that take too long to convey a meaning or convey none at all. Deadwood belongs in the second category, for, as the name suggests, deadwood is completely unnecessary words and phrases that somehow slip into writing. Because they are unnecessary to the meaning of a sentence, such expressions can simply be dropped out of the sentence without loss, and they require no more labor during revision than simply marking them out of the draft. Each of the following examples contains deadwood, and the improved versions show that with the unnecessary words and phrases eliminated, the sentences have suffered no loss of meaning. Moreover, the actual messages come through a bit more clearly after the deadwood has been removed, because their statement is less cluttered than before.

1. A portion of the signal can be shorted to ground by lowered insulation resistance at a clamp or clip-*type holding device*.

   Improved: A portion of the signal can be shorted to ground by lowered insulation resistance at a clamp or clip.
2. The capacitance change will vary according to the temperature coefficient of the basic dielectric *which was* used in *the construction of* the capacitor.

   Improved: The capacitance change will vary according to the temperature coefficient of the basic dielectric used in the capacitor.
3. This organization must obtain new design disclosure data *as available* and eliminate obsolete or nonpertinent data *no longer needed*.

   Improved: This organization must obtain new design disclosure data and eliminate obsolete or nonpertinent data.

Often, the use of deadwood is a characteristic of ordinary conversation. In talking about their ideas, people can say, unnecessarily, that the sky is blue *in color* or that the building is large *in size*, but their listeners are not likely to pay much attention to these unnecessary qualifications. In writing, however, readers are much more likely to begrudge the time that they spend on a reminder, for example, that blue is a color. Consequently, persons who tend to write as they talk should especially be on the lookout for deadwood when they revise their written material.

Writers also should realize that although they may eliminate only one or two short words from a given sentence, each small economy in language contributes to the conciseness of an entire report. Deadwood and other kinds of wordiness do not tend to be concentrated in a few passages. Rather, they tend to be scattered throughout many sentences. Thus writers usually achieve conciseness by saving a syllable or two in one sentence, a word in the next sentence, two or three words in the third, and so on.

Another cause of wordiness is the use of roundabout expressions or expressions that require several words to convey an idea when one or two would serve as well. In the following sentences, for example, the italicized phrases take longer than necessary to say what must be said.

1. He wrote me *in regard to* your proposal to reduce the clerical cost *in connection with* auditing travel vouchers.
2. This work, divided between current and *long-range planning type* vehicles, *provides an area for* the practicing engineer to display his imagination and knowledge.

In the first sentence, the phrases *in regard to* and *in connection with* come from a large stock of wordy expressions that are available in English and are all too commonly used. Indeed, such expressions become habitual with a writer, and locating them in a draft can require real effort. The roundabout phrases of the second sentence, on the other hand, probably resulted from the lack of precision that often shows up in a first draft. Unfortunately, it seems, the writer paid too little attention during revision to improving the draft's clarity and conciseness. In both sentences, the writers could have reduced their wordiness by substituting a single word for each of their roundabout phrases.

1. He wrote to me *about* your proposal to reduce the clerical cost *of* auditing travel vouchers.
2. This work, divided between current and *future* vehicles, *permits* the practicing engineer to display his imagination and knowledge.

A third practice that can cost words is the overuse of indirect openings for sentences—*there is, there are, it is*. Very probably, writers will not care to revise all their sentences that begin with such expressions. Sometimes, the indirect opening may very well be the best way to get a sentence started; at other times, this type of opening provides a welcome change of sentence structure. Nevertheless, beginning a sentence with *it is,* for example, delays the introduction of the sentence's subject, and the inevitable result is extra, if not excessive, wordage.

Consequently, writers who often use indirect openings should examine the sentences in which they have done so. More often than not, they should decide to undertake the kind of revision suggested below:

1. *There are* many factors that affect the construction of a dam.

   Improved: Many factors affect the construction of a dam.

2. *It is* this phenomenon of dielectric absorption that accounts for the fact that a capacitor discharged by a short circuit often has a small charge after the short is removed.

   Improved: Dielectric absorption is the reason that a capacitor discharged by a short circuit has a small charge after a short is removed.

Effective revision of the second sentence, incidentally, not only would eliminate the indirect opening but also would eliminate the deadwood, *this phenomenon of . . . that,* and would substitute a shorter phrase for the roundabout *accounts for the fact.* Curiously enough, indirect openings have an unfortunate way of attracting other types of wordiness.

Conciseness in writing is important to effective communication, but if writers are to achieve conciseness, they must do so by looking for words and phrases that contribute little or nothing to meaning. Deadwood, the roundabout expression, and the indirect opening are three of the most common weaknesses that prevent conciseness in much technical and scientific writing. Yet they are easy to eliminate if writers would but first identify them in a draft and then take the pains to make the changes required.

Listing Detail. The listing of details within sentences can give readers very much the same kind of difficulty as can the listing of ideas in the sentences of a paragraph. At the sentence level, listing takes the form of conveying an idea bit by bit in prepositional phrase after prepositional phrase. A good example of the practice follows:

(In recognition) (of the need) (for information) (of importance) (in the evaluation) (of the hazards) (of nuclear systems) intended (for auxiliary power use) the U.S. Atomic Energy Commission has initiated aerospace safety studies (as part) (of its overall nuclear safety programs) (under the nuclear safety and engineering test branch) (of the division) (of reactor development).

In this sentence, most of the details making up the statement of an idea are parceled out among the eight prepositional phrases that precede the subject *(Commission)* and the verb *(has initiated)* and then among five more such phrases that follow. The result is a kind of density that readers have trouble in penetrating. Because each detail is presented in a similar structure, each one takes on a role of equal importance. Thus readers must sort over the list to establish the full coordinate and subordinate relationships that presumably exist among all the details the sentence offers. Usually, sentences like this one will not stop a reader's progress through a report, but they can slow it down considerably. Moreover, their frequency in technical and scientific writing has helped, unfortunately, to make it more difficult and dull for readers than it need be.

Some readers of the sentence being discussed here may note that it is rather long and, perhaps, can see its length as being the cause of their problems with it. And, indeed, it is longer than most sentences should be in writing. No doubt, also, reorganizing it into two or more sentences could improve its readability. As it stands, however, the sentence "reads longer" than many other sentences that contain more words but fewer prepositional phrases. Simply revising long sentences into shorter ones full of these phrases can be of less help to readers than writers might believe.

To avoid the listing of detail in their sentences, writers should use more subordinate clauses and fewer prepositional phrases. By using subordinate clauses, writers can actually do some sorting out of a sentence's details and can clearly and firmly express important relationships among them. The resulting sentence will be less dense and more readable than one in which details are merely listed. Below, a revised version of the example illustrates how the use of the subordinate clause, as well as the use of two sentences, can increase the original's readability.

The U.S. Atomic Energy Commission recognizes that important information is needed for evaluating the hazards of nuclear systems which may be used for auxiliary power. Consequently, it has initiated aerospace safety studies which, as part of its overall nuclear safety programs, are directed by the nuclear safety and engineering test branch of the reactor development division.

Too often, density in technical writing results from inadequate style rather than from the supposed complexity of the material being presented. Writers who work in revision to eliminate listing from their sentences find that frequently they can present highly complex material clearly and readably.

Words

The proper choice of words, of course, is vital to stating ideas effectively, and choosing the right word is not always easy. It requires not only that writers know what they want to say, but also that they have an awareness of the meaning of words and the subtle differences that can exist among those meanings.

During revision, writers should carefully examine the words they actually did choose while writing their drafts and should replace any that misrepresent, obscure, or weaken their messages. They need to make sure that they have used each word accurately and that it truly represents what they want to say. They also should beware of using unnecessarily "big" words that, at worst, can cause their styles to be obscure and, at best, can sound pompous or even "phoney." Finally, they should do what they honestly can about unstacking the hordes of modifiers which seem to overwhelm so many nouns in technical writing.

Inaccurate Words. In reconsidering the accuracy of words they have used in a draft, writers would do best to play the role of the complete skeptic. They should assume that they have used inaccurately any word about which they can have the slightest doubt, whether it be an old, familiar word they are using in a new way or a new word recently acquired. Whatever the reason for the doubt, the next step, of course, is to resolve it by consulting a good dictionary and either confirming that the word is the appropriate one or finding a substitute that is accurate.

Such advice may sound overly cautious to many writers, and, for some, perhaps it is. Nevertheless, caution would have prevented three writers from letting the following mistakes get into print.

1. The Columbus Day storm and subsequent storms *incurred* heavy damage to the timber resource.
2. Table I *references* the environmental limits established for the missile system.
3. Where widely *divergent* temperatures are expected, ...

In the first example, the writer may have mistaken *incurred* for the word *inflicted*. Obviously, though, the storms did not bring damage upon themselves; they inflicted damage on the timber resource. In addition, readers may wonder why the writer did not simply say that the storms heavily damaged the timber resource. So far as the second example is concerned, dictionary makers have been slow to accept that the noun *references* may properly be used as a verb. Moreover, the table mentioned in the sentence did not refer the reader to the environmental limits; it listed them. In the final example, the writer did not really mean *divergent temperatures,* as two temperature curves might diverge from a more or less common point on a chart. He meant *widely different temperatures*. And a quick trip to the dictionary for some advice on the possible synonyms *differ* and *diverge* (perhaps *different* and *divergent*) would have made clear the difference in meanings of the two words. Indeed, a general word of caution may be in order here: as rich as the vocabulary of the English language is, it contains relatively few exact synonyms.

Although engineers and scientists give high priority to handling the tools and instruments of their work with great care, many of them, unfortunately, do not give the same care to their use of language and especially their choice of words. Yet the tools of language are essential to the effectiveness of what they say. Effective communication must be accurate, and accuracy depends a great deal on the choice of the right word. Writers, then, should revise their drafts carefully to make certain that they have used the right words to express the meanings they wish to convey. If inaccurate use of words does not impede understanding, it usually makes a statement look silly.

Unnecessarily "Big" Words. The "big" (unusual or uncommon) word and the "little" (more common, often shorter) word are neither good nor bad in themselves. The real value of a word is determined by how well it states what a writer wishes to say. Yet, in technical and scientific writing, the rule often seems to be that the "bigger" a word is, the more likely writers are to choose it. So widespread is the overuse of *unnecessarily* "big" words that many organizations have compiled lists of those that appear everywhere in their reports and correspondence. The lists are always long, but their compilers never manage to keep up with the latest fashions that spread among technical writers in no time at all.

Curious readers can wonder how long it has been since an engineer or scientist *used* something rather than *utilized* it. Or when the last time was that an organization *started* and *ended* a project instead of *initiating* and *terminating* it. Of course, readers never expect to learn that plans for a project have been *completed*; plans are always *finalized*. On the other hand, they can expect to encounter *facilitate, ascertain, implement, cognizant, procure, deem, fabricate,* and hundreds of other "big" words that help to make up the vocabulary of far too many sentences in technical writing.

Explaining the widespread use of these unnecessarily "big" words is difficult. It is important for writers to recognize, however, that "big" words usually do not communicate any more accurately or effectively than simpler substitutes. In fact, they can seriously obscure meaning. Moreover, they do not impress most readers with how learned the writer is. In fact, they frequently convey a pretentious tone that readers find either amusing or irritating. Truly effective writers, writers who impress their readers, are those who state their ideas clearly and simply, using words no more difficult than necessary to convey their ideas.

Although the "big" words being discussed here are widely used in technical writing, they have no claim to being technical terms. Certainly, if such terms must be used to communicate clearly, then writers should use them. However, writers also should include definitions of their technical terminology if they have any questions about the ability of their readers to understand it. Writers should remember how they resist consulting a dictionary for the meaning of words that they do not understand when they are reading, and they should recognize that their own readers are not anxious to use a dictionary any more often than they are. If writers wish to be understood readily, they should choose and handle their words with some thought for the reader's needs. More often than not, the simple word serves the reader best.

Stacked Modifiers. The listing of details can take place within parts of the sentence, just as it can within the structure of the sentence as a whole. It occurs every time a writer modifies a noun with a long string of adjectives and other nouns, and produces what is often called stacked modifiers. A good example of the result appears at the start of the following sentence: "*The maximum permissible aisle seat speech interference* level is 70 db." In this case, the stacked modifiers, stretching from *The* through *interference,* make up more than half the sentence's length. They tend to bury the noun *level,* which they are supposed to modify in some fashion, and they are little more than a list of qualifying details in which readers must find the pattern of relationships.

Unfortunately, too many technical writers see the stacking of modifiers as a way to clear and concise writing—clear because they read into each stack of words relationships that they understand, and concise because stacking is the briefest form a group of modifiers can take. Readers, however, see stacked modifiers in a very different light. This kind of listing confronts them with a density of expression which equals that produced by the overuse of prepositional phrases. Each set of stacked modifiers forces the reader to sort out the various modifiers and then to determine how they relate to one another and to the noun being modified. This task is not always simple, as the following examples illustrate.

1. This program includes *long-term surveillance test* planning.
2. The *heterogeneous graphite moderated natural uranium fueled low-powered critical nuclear* reactor. . . .

Although in the first example only three words modify *planning,* readers can waste considerable time and thought while trying to determine whether the writer meant long-term planning of surveillance tests or planning of long-term surveillance tests. Nothing in the sentence indicates which alternative the reader should choose, and the context from which the sentence came was equally unhelpful. The second example is, of course, much more complex and technical than the first. Nevertheless, the reader comes down to the same question: What does the phrase mean? Actually, the colleagues of the man who wrote it spent two hours arguing about its meaning. Then they concluded that it did not accurately convey the nature of the reactor.

What can writers do about stacked modifiers during revision? There are a number of possible means for reducing their density. Sometimes, a comma inserted betwen two modifiers will help the reader to understand their relationship. At other times, a well-placed hyphen will help. For example, the writer of the sentence on test planning could have revised his stacked modifiers to read, "long-term-surveillance test planning," and his audience probably would have understood what he really intended to say. Often, however, the best approach is to reorganize sentences in which stacked modifiers appear and to place some of the modifiers, at least, in prepositonal phrases or subordinate clauses. Doing so can make it unnecessary for readers to work out for themselves the relationships among the modifiers, as the following examples illustrate.

1. *The batch and continuous dissolution and solvent extraction* flowsheets embody the principal results of the study.

   Improved: Flowsheets for both batch and continuous dissolution and for solvent extraction embody the principal results of the study.

2. . . . *six one-half inch diameter by .002 inch maximum thick* foils.

   Improved: . . . six foils which are one-half inch in diameter and have a maximum thickness of .002 inch.

These examples also show, of course, that in revising to reduce the density of their statements, writers must frequently use more words than they had in their drafts.

If a report contains a number of the stylistic weaknesses discussed in this chapter, the chances are that its language will resemble something called jargon, a term popularized by Sir Arthur Quiller-Couch to label verbal fuzziness. Jargon is never really acceptable English. At best, its meanings are seriously blurred. At worst, it conveys no meaning at all. Such was the not unusual discovery of a manager who received a progress report beginning with the following paragraph.

> Determining whether or not a human crew is necessary will follow from a system requirement for the origination of new commands directing the guided component against one or more from a large collection of emerging objectives, with varying probability of occurrence of objectives. How large is "large" will be determined in this study.

The manager still does not know what this paragraph (and the three that followed it) meant to say. Fortunately, circumstances made it unnecessary for him to explain to someone else what one of his staff members was doing. Otherwise, he could have wished that the writer had not been so addicted to "big," abstract words listed in prepositional phrases.

Not too long ago, an article in a Pacific Northwest newspaper attributed to Vice Admiral Hyman A. Rickover the following comments about a report he had then recently received.

> This acknowledges receipt of your memorandum and attachment requesting my review and comments on your *Guide for the Preparation of Special Analytical Studies*. I have spent much time reading this document; unfortunately, I cannot understand it. Many statement are beyond my comprehension; for example:
>
> > The concept of a parallel internal list of topics in addition to those which are specifically identified for near-term submission to the D.O.D. recognizes an agency need or interest for initiation of study activity in areas in which it is not clear prior to completion that discussion with D.O.D. will be warranted, or which may represent possible early phase of more formal studies later or which may require an extended period for completion.
>
> I have asked several of my leading scientists and engineers to help me, but they also found your Guide beyond their comprehension. My conclusion is that we in Naval Reactors are not sufficiently sophisticated to understand it; in order to ascertain if your Guide has any practical use, it would first have to be rewritten in simple English.

What more need be said?

Perhaps a quotation from *The Elements of Style* sums up the principle that writers should adopt in all of their work while writing and revising. In a passage on conciseness, William Strunk, Jr., closes with a phrase as memorable as any in his "little book":

> Vigorous writing is concise. A sentence should contain no unnecessary words, a paragraph no unnecessary sentences, for the same reason that a drawing should have no unnecessary lines and a machine no unnecessary parts. This requires not that the writer make all his sentences short, or that he avoid all detail and treat his subjects only in outline, but that every word tell.[3]

"That every word tell" should be the aim of every report writer.

[3]William Strunk, Jr., and E.B. White, *The Elements of Style*, The Macmillan Company, New York, © 1959; paperback edition, 1962, p. 17.

EXERCISES AND DISCUSSION TOPICS

1. Identify the factors that cause you difficulties in getting started at writing a first draft. Think of practical ways in which you might overcome these difficulties.

2. Exchange the draft of a report or paper with a fellow student and suggest revisions for each other's written work. Be sure that each of you has a clear understanding of the purpose and audience of the report or paper, and consider the effectiveness of the draft's content, organization, style, tone, and (if

appropriate) use of illustrations and layout. Be tactful in making your suggestions, and *listen* to what the other student has to say about your draft.

3. Collect as many examples as you can find of inflated tone and inappropriate use of the impersonal and negative tones in writing. Sources should be everywhere: textbooks; newspapers; magazines; reports, letters, and memos of all kinds—anything in print.
4. Choose one (or more) of the examples of poor tone that you have collected and describe how the writer's use of language produced the resulting tone. Revise the example (or a passage from it) to improve the tone.
5. Collect as many examples as you can find of writing in which the style and tone seem particularly well suited to the writer's purpose and audience.
6. Choose one (or more) of the good examples of style and tone that you have collected and describe how the writer produced the effective result that he or she did.
7. Revise the following paragraphs to eliminate the listing of ideas (''primer'' style):
   a. Throughout the month we operated with a 10 percent reduction in personnel. There are only 25 unanswered letters in this office today.
   b. The other piece is 5/16 in. thick. It has a rectangular shape. It is 5¾ in. long and 1¾ in. wide. It is made from a plastic material.
   c. Many solutions have been proposed for the problem. The auxiliary power plant is now widely used. This is because it supplies power when needed. It can also be turned off when not needed.
   d. The pressure increased rapidly. The pressure pushed against the sides of the tank. It increased until it was very high. At last it increased to the point where the tank blew up.
   e. Resistors having molded plastic around the resistive elements are affected to a minor extent by pressure changes. The direct effects of pressure are negligible. The combined effects of pressure and humidity can be severe upon a molded resistor's electrical characteristics. Pressure changes coincident with the presence of humidity cause ''breathing'' of the part. Moisture is absorbed into the body material and the insulation resistance is lowered. The amount of moisture absorbed depends upon the moisture absorption characteristics of the body material.
8. Revise the following sentences to correct any grammatical faults that may be present, to increase their conciseness whenever possible, and generally, to improve their simplicity, accuracy, and clarity of statement:
   a. The only way to correct a bad habit is to substitute it for a good one.
   b. This location was satisfactory when the work was scattered throughout the division. But now that it is concentrated at Longview it is unsatisfactory.
   c. When working on a large scale, these conditions present problems difficult to cope with.
   d. If there are further questions which you may have in regard to your duties with our group or living and working conditions here in Portland, please feel free to write to me.
   e. This model developed an intermittent ice cap that collected in the vee cup of the tip at a total power output of 525 watts.
   f. The Thermogun should be used in lieu of the Thermofit tool.
   g. Review that requirements are clearly and simply stated.
   h. It is a foregone conclusion that the future use of epoxy resins in piping is assured as a product of progress by the growing recognition shown it by engineers and engineering companies.
   i. To demonstrate that such waves are virtually invisible at sea, the captain of the ship standing off the port of Hilo, Hawaii, was astonished as he

watched the harbor and much of the city being demolished by waves he had not noticed passing under his ship.

j. The committee will review the production test plan as published against agreements reached in the meeting before approval.

k. This question, which was asked to the men in management at Smith and Jones Company, was answered in many ways.

l. Drifts occur as a result of one turn of wire shorting to an adjacent turn, thus reducing the resistance.

m. In most capacitors, the phenomenon known as dielectric strength is present in varying quantities.

n. After follow-up has affected complete resolution, approve the document for submission to the customer.

o. It is sometimes necessary to request that work be performed during scheduled plant shutdowns. This should be put in a memo form that is sent to the scheduling supervisor only.

p. Check the documentation to see that no redundant measurements are called out.

q. The bonds seem to be both chemical and mechanical in nature.

r. The most significant area for investigation involves devices to reduce impact severity of the driver and passengers against interior protrusions and protection against complete ejection from the vehicle.

s. It has been established through subsequent research that this system for manufacturing clay pipe fittings will replace the old method of fabrication.

t. The resonances of the part and the resonances of the mounting structure must be apart.

u. Initiate follow-up to ensure completion of corrective action.

v. It is recommended that the project engineer be cognizant of testing equipment conditions and when found to be inferior make arrangements for its repair or replacement.

w. It is intended that adherence to the format as outlined will permit work, through a critical and analytical approach, to be effectively carried out to completion.

x. The insulation resistance is usually in the order of thousands of megohms at 20°C.

CHAPTER 7

Abstracts and Summaries

For many persons, the terms *abstract* and *summary* have become interchangeable; yet an important distinction between them has grown out of actual practice in industry and government. More and more often these days, the term *abstract* refers to a brief, descriptive statement of what a report or journal article contains. Like a title or a table of contents, the abstract simply indicates what the coverage and content are. On the other hand, the term *summary* means a short condensation of a report's content. The summary is informative rather than descriptive, for it actually contains the most significant facts and judgments presented in the report.

Part of the confusion between abstracts and summaries has sprung from the practice of dividing summaries into two classes: descriptive summaries and informative summaries. Although both types of statements are to be found in reports, it would help to clarify matters if descriptive summaries were called abstracts and the term *summary* were confined to informative summaries. This simple adjustment would mean that a pair of common terms could be used to talk about actual practice in scientific literature and in engineering reports. It also would retain the useful distinction between abstract and summary, which is a part of scientific and technical writing.

The (descriptive) abstract and the (informative) summary are important communication tools, each with its own purpose and value for readers. Although both come at the beginning of a report or an article so that the reader can read them first, it is best to write them after the report or article has been finished.

ABSTRACTS

The purpose of the abstract is to tell potential readers what kind of information they will find in a report or an article. If the abstract is written well, it is a concise, but accurate, description of what the report or article takes up in detail. Thus it can help readers to decide whether they should spend the time and effort to read any further. Frequently, to extend the range of such help, abstracts are included in the entries of library files and in published indexes. In this case, however, potential readers can decide on how useful a report or an article may be to them without ever seeing it.

To be helpful, an abstract must describe the content of a report or an article with enough care so that readers can make accurate judgments about it. The abstract must make clear what the exact nature and scope of the subject matter are. If special procedures, tests, or equipment were used in the work being reported on, the abstract should note this fact. It should state the nature and show the order of the ideas to be developed, and it should indicate what is emphasized.

Meeting the requirements of this recipe in a paragraph or two can take some time and care. Yet, in today's flood of scientific and technical information, the specialist reader needs the convenient shortcut to identifying useful information which abstracts like the following example represent.

ABSTRACT

Correlation and spectral functions resulting from the analysis of organic phase volume fraction measurements on a pulsed, sieve-plate solvent extraction column are presented, and some of the changes that occur as the flow characteristics of the column change are shown. Spectral functions reveal the shifts in operation more directly than do the correlation functions. General trends of both correlation and spectral functions with changing flow characteristics are discussed. Spectral density ratios which reflect the power density in the low frequency bandwidth from 0 to 0.1 Hz relative to that at the pulsing frequency are defined. These ratios are shown to correlate directly with the mode of column operation and are especially sensitive to incipient flooding. Hard limiting of the data prior to spectral analysis is shown to have little detrimental effect on the calculated functions of interest.[1]

SUMMARIES

The informative summary, the short condensation of the essential content of a piece of writing, is normally found with a report rather than an article. It usually is placed "up front" where readers cannot miss it.

For many readers, the summary *is* the report—that part of the report they do read, anyway. Managers, for example, frequently depend on the summary for their information, especially if a report is long or complex. Consequently, effective and truly informative summaries are vital to communication, and scientific and technical writers need to develop skills in writing them.

Writers have two approaches available to them in preparing summaries: (1) to summarize briefly *each major portion* of the report or (2) to summarize the *most essential content* of the report. For example, a summary of the entire report on the solution of a problem would include a brief description of the problem, the approach taken to solve it, and the results and final conclusions. A summary emphasizing the essential or most important content, on the other hand, would concentrate on the problem and its solution, but would omit discussion of how the solution was obtained. Most often, these selective summaries emphasize solution, conclusions, recommendations.

If an informative summary is well written, its readers learn what the significant content of a report is without reading each and every section. They are able to discuss, for example, the nature of a research project, its findings, and the recommendations that may have grown out of the work. At times, management decisions are based on what a summary conveys; thus writers must take care to provide not only a brief but also an undistorted view of a report's main ideas. For this reason, informative summaries are usually longer and more detailed than descriptive abstracts. Although abstracts are usually limited to one or two paragraphs, summaries frequently require one or two pages, possibly even four or five for long and complex reports. Summaries should be short, of course, but they also must be complete enough to serve their purpose. The following summary is an example of an informative one, which really tells the reader what a report has to say.

GAS FLOWMETERS

The object of this experiment was to measure and to study the rate and characteristics of the flow of air through the following flowmeters: a venturi tube, a thermal flowmeter, a Pitot tube, an orifice, a nozzle, and a short tube orifice. The apparatus consisted of a 6-inch fabricated pipe loop with the meters arranged in the sequence above. A positive-displacement blower circulated the air at 2 psig, and meter readings were taken at various conditions of flow, the velocities being controlled and adjusted to give good representative readings. The thermal flowmeter was set at 650 w, a stipulated condition.

Experimental errors were confined to meter readers; they were estimated at 0.5 percent for pressure and 0.1 percent for temperature readings.

Certain considerations were neglected or approximated, such as vena contracta friction losses, jet contraction, density and velocity changes, ideal gas law deviations, and pulsating flow. These unknown variables were corrected by experimentally determining the average coefficient of discharge C, the measure of departure from ideality. The venturi tube was employed as the standard device to calibrate the other meters. With the venturi C = 0.98, the rate of flow W was found to be 0.40 lb/sec. The C's for the Pitot tube, the orifice, the nozzle, and the short tube were calculated to be 0.87, 0.46, 0.89, and 0.87, respectively. These values compared within 5 percent of literature data in all flowmeters except the orifice (0.46 versus 0.61 by literature). Permanent pressure loss compared favorably with accepted values. The thermal flowmeter was used to check the venturi C by incorporating its data in an energy balance; on this basis, C equalled 0.93.

The following conclusions may be drawn. The venturi tube, at low permanent pressure loss and high cost, was the most efficient device tested. The nozzle, the Pitot tube, the short tube, and orifice, progressively in that order, increase

[1]Michael J. Strand and Kermit L. Garlid, "The Application of Spectral Analysis in Monitoring Extraction Column Behavior," *Canad. J. Chem. Engineering,* Vol. 53, December, 1975, p. 677.

in permanent pressure losses. The thermal flowmeter is accurate at moderate air velocities at which low heat losses occur.

In preparing summaries, writers should first make a list of the main ideas presented in their reports. Once they have identified the content for their summaries, they should start to write, making certain that they

- Preserve the order of the ideas in the report itself, except for those sections not included in the summary.
- Include no information not part of the report.
- Provide continuity by relating the main ideas so that the summary has a unified impact.

At times, writers may have to provide both a descriptive abstract and an informative summary for something that they have written. But such occasions are rare. Much more often than not, they must prepare one or the other, and someone else has already decided which one it should be. When they don't have a choice, writers should consider the different functions that the abstract and summary have and provide the one that will serve their readers best.

EXERCISES AND DISCUSSION TOPICS

1. Write a descriptive abstract for your report. Write an informative summary for it.

2. Select and read a magazine article. Write a descriptive abstract for it. Write an informative summary for it.

Epilogue

In closing this discussion of scientific and technical writing, we would like to present two additional ideas, which we think are important not only for writers but for their organizations as well. The first is that there is a point of diminishing returns in writing, after which the amount of time expended to gain a higher degree of quality is too great for the results obtained. The second is that all scientific and technical writing need not possess the same degree of quality.

As obvious as these ideas may be, too often they are overlooked by professionals faced with the task of writing. Many times, of course, these writers are at a loss on how to measure time versus quality in any realistic sense, and a balance between these two factors is, therefore, difficult for them to achieve. Deadlines, as realisitic or unrealistic as they may be, usually force a solution on the writer. Yet a little thoughful sharing of ideas can produce a reasonable estimate of where the balance between time and quality should lie for any given type of writing.

Writing, like any other human activity, does possess what we can call an "improvement curve," and although we have not measured that curve accurately or scientifically, we know its general shape. If we plot the time spent versus degree of quality achieved, we find that the curve ascends steeply at first, then starts to flatten out, until it reaches a peak, and then most likely drops back down to some extent at its end. We have represented what we believe this curve would be like in Figure 19.

Because perfection is never really achieved, the curve simply does not reach the top of the chart. Moreover, if we work long enough on a piece of writing, we are very likely to reduce its quality a bit by changing earlier modifications that were actually better than the new ones; thus the curve drops off some after its peak.

Understanding the shape of the improvement curve can assist scientific and technical writers, as well as their supervisors, in two very distinct ways. First, the curve shows that simply doubling the amount of time devoted to a piece of writing does not ensure that the quality of that writing will improve in any similar ratio. The shape of the curve suggests that for earlier periods of time spent in writing, a great deal of quality change results. However, it also suggests that as more and more time is devoted to writing, less and less quality will be obtained. What is needed, of course, is to develop a realistic sense of where we are in relationship to this curve on any given project. All writing takes place in some kind of time frame, and our task is to bring the time we expend into balance with the

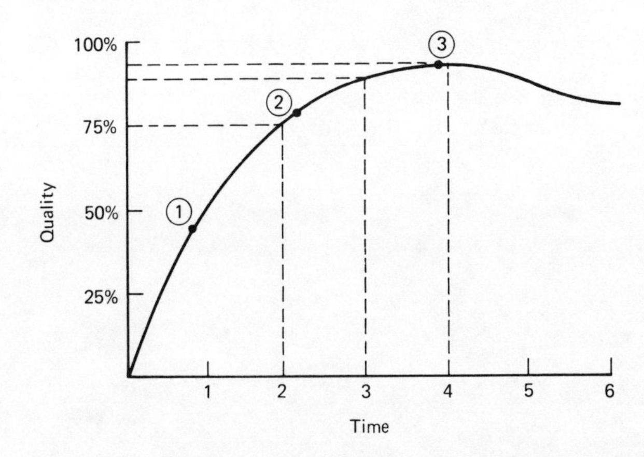

Figure 19. The writing improvement curve can help writers in two ways: (1) to realize that they do not get the same amount of improvement for every hour they spend writing or revising and (2) to recognize that all of the writing they turn out need not be at the same level of quality. For example, a letter requesting personnel might be at one, a budget request at two, and a professional journal article at three.

degree of quality we want for a given piece of writing. Understanding the improvement curve is essential to achieving such a balance.

The second use of the improvement curve that would be extremely beneficial both to writers and their supervisors involves a definition of quality standards for the various kinds of writng produced. Writers need to know the level of quality expected of them for the various kinds of writing produced. Writers need to many people assume that all writing must be of the same quality. But in reality, aren't some of the kinds of writing we produce quite adequate even if they are ''quick and dirty''? Some of the things we write must, of course, possess as high a quality as we can provide.

Many of the kinds of writing that we do, however, belong somewhere between these two extremes. It would be helpful if technical and scientific writers and their supervisors would discuss the degree of quality required for the different kinds of writing that the staff must produce. Definition of quality standards would do much to clarify the writing task, both for the writer and for the reviewing supervisor.

Writing is varied. A memo to the shop, an order for parts from the storeroom, a memo to the personnel department requesting a technician, an article for a learned journal, or an environmental impact statement to come under the scrutiny of a concerned public—all require a different degree of polish and quality. Certainly each of these kinds of writing exists in an environment of its own. Because the use requirements vary in each environment, we suggest that the quality of writing might also vary.

We are not suggesting, of course, that writing can be unclear, ambiguous, distorted, or sloppy. What we are suggesting is that the time and effort writers and their reviewing supervisors put into seeking the most effective way of stating an idea should not be the same for all kinds of writing. We would place the various kinds of writing performed by the staff in any organization at different points along the improvement curve so that we could establish standards for quality that are related realistically to the different writing situations. In doing so, we could establish the general standards of quality to be expected within the organization for the variety of writing it produces, and, in most organizations, such an action might well be revolutionary.

Even though this improvement curve is not a precise tool, it provides a basis for considering the time frame for a piece of writing. Each of us needs to develop a sense of the relationships of quality versus time and to recognize their significance for the kinds of writing we must produce and the writing standards we must meet.

Selected Bibliography

Many excellent books on technical communciation are available, and those of us who are frequently involved in technical writing need to take advantage of the help they can give. The following bibliography lists volumes that we ourselves have found especially useful.

HUMAN COMMUNCIATIONS SERIES

Wiley-InterScience publishes a valuable series of books on technical communication. Its wide variety of titles includes treatment of a number of specialized subjects not available elsewhere.

William Bowman, *Graphic Communication*

An excellent and useful book on principles of graphic design, well organized and well illustrated. This work is a valuable starting point for the would-be illustrator or for the writer-editor who wants to better understand graphic design.

George A. Magnan, *Using Technical Art*

A practical book that covers drawings, photographs, and visuals for presentations. Effectively presents the alternatives in the use of technical art on a "which will communicate?" basis. Well illustrated. A required book for writers and editors.

Milton Feinberg, *Techniques of Photo Journalism*

For the person who wants to know how to take pictures with quality and impact. From basic equipment to specific assignments, this book keeps a focus on photographic principles. It features helpful and practical do's and don'ts. An excellent book for the serious photographer.

Jordan, Kleinman, Shimberg, *Handbook of Technical Writing Practices*, 2 volumes

These two volumes make up the basic library for professional writers, editors, and publications managers. Produced in cooperation with the Society for Technical Communciation, they present a valuable resource covering all aspects of technical publications, from the various kinds of documents and service necessary to their publication to managing a publications unit.

Vardaman, Halterman, *Managerial Control Through Communication*

This book on management stresses the role of communication in the diagnosis, design, and management of organizations. Includes a number of selections and reprints of articles on communciation and organization.

Robert H. Dodds, *Writing for Technical and Business Magazines*

Full of practical suggestions, this book focuses on the publishing process. It not only covers the author's problems but also relates them to those of the editor and the publishing process. Draws on the input from many magazine editors. Down-to-earth and practical. Good for students and professionals alike.

Michael P. Jaquish, *Personal Resume Preparation*

A good "little" book on resume writing. Good attention to specific problems and useful examples.

Herman M. Weisman, *Technical Correspondence*

A useful handbook and reference for writing technical correspondence. Presents and applies basic principles. Good coverage of various types of letters with examples. Helpful for anyone who writes technical correspondence.

Stormes and Crumper, *Television Communications Systems for Business and Industry*

A helpful book for those who want to know how closed-circuit television works, what problems it can solve, and what its limitations are. Recent advances in equipment must be added to the basic presentation in the book.

John H. Mitchell, *Writing for Professional and Technical Journals*

This work discusses such writing problems as organization, key words, usage, documentation, bibliography, and illustrations. It also presents sample style guides for articles from a broad scope of professional journals in scientific, technical, social science, and humanities fields.

John B. Bennett, *Editing for Engineers*

A useful presentation of principles and techniques for the engineer who must review reports from staff members. Helpful not only to such managers but to writers who are serious about quality writing and who are willing to revise in order to achieve effectiveness.

John D. Haughney, *Effective Catalogs*

A valuable book for those responsible for developing catalog programs for industry. Full of useful, practical information on planning; cost analysis; catalog design; and production, distribution, and marketing coordination.

James J. Welsh, *The Speech Writing Guide*

A useful work full of practical advice on speech preparation and presentation, including visual aids. Convenient checklists and list of information sources add to its usefulness.

W.A. Mambert, *Presenting Technical Ideas*

An excellent, highly practical, highly readable book. Drawn from real-life situations, it combines practical experience and proven academic principles. Should be on the desk of everyone who must prepare and deliver technical presentations.

Harold E. Daubert, *Industrial Publicity*

A highly practical and comprehensive book that relates industrial publicity to public relations, marketing, and advertising. Its emphasis on planning, producing, and evaluating publicity combined with its useful checklists makes it an excellent guide and reference.

GENERAL BOOKS ON TECHNICAL WRITING

Bostwick, Burdette E., *Resume Writing*, John Wiley & Sons, New York, 1976.

A useful, practical guide for those who must write resumes. Its comprehensive coverage of 10 resume styles and forms provides the reader with a flexible and varied attack in job seeking. Many useful examples.

Ehrlich, Eugene, and Daniel Murphy, *The Art of Technical Writing*, Apollo edition, Thomas Y. Crowell, New York, 1969.

A useful paperback that takes up a variety of forms of technical writing, including the proposal and journal article. Also includes a chapter on the management of technical writing and a handbook of style and usage.

Estrin, Herman A., *Technical and Professional Writing, A Practical Anthology*, Harcourt, Brace and World, Inc., New York, 1963.

An interesting collection of articles and papers on technical writing. Valuable for the insights from many persons with quite varied experience.

Ewing, David W., *Writing for Results, in Business, Government, and the Professions*, Wiley-Interscience, John Wiley & Sons, New York, 1974.

An interesting, down-to-earth, useful book by the executive editor of the *Harvard Business Review*. Its emphasis on writer-reader relationships and communciation strategy provides valuable insights. Its standards for evaluating writing both in content and expression make it highly useful and instructive.

Gallagher, William J., *Report Writing for Management*, Addison-Wesley Publishing Company, Reading, Mass., 1969.

A book that gives very sound, practical advice to persons who must write reports to management. It emphasizes the interaction of writer, reviewer, and reader. Useful not only for writers but also for managers who are responsible for reviewing reports.

Hicks, Tyler G., *Successful Technical Writing*, McGraw-Hill, New York, 1959.

Covers a variety of different types of technical writing: articles, papers, reports, instruction and training manuals, and books. Treatment of reports is sketchy, but the other types of writing are treated fully. One of the few sources of information on writing manuals and books. Treats articles in detail.

Houp, Kenneth W., and Thomas E. Pearsall, *Reporting Technical Information*, second edition, Glencoe Press, New York, 1973.

An excellent and widely used textbook. Provides an effective overview of technical writing, with particular emphasis on analyzing the audience. It takes up graphical elements and a wide variety of kinds of technical writing.

Johnson, Thomas P., *Analytical Writing: A Handbook for Business and Technical Writers*, Harper & Row, New York, 1966. (Presently out of print.)

One of the best books on how to diagnose and eliminate the faults of style found in much technical writing. Analysis of many examples, good and bad. Also advocates using the "inverted pyramid" approach to organization. A book the writer can use for self-improvement.

Kapp, Reginald D., *The Presentation of Technical Information*, revised edition, Constable, London, 1973.

An excellent book from England. Discusses the problems of technical communication at a more philosophical level than do most others. A book that should be read by every writer of reports. Discussion of "Do's and Don'ts." Good examples.

Mills, Gordon H., and John A. Walters, *Technical Writing*, third editon. Holt, Rinehart, and Winston, New York, 1970.

A popular and widely used textbook. Good material on special technical writing problems; on writing transitions, introductions, and conclusions; on types of reports; and on layout.

Sherman, Theodore A., and Simon S. Johnson, *Modern Technical Writing*, third edition, Prentice-Hall, Englewood Cliffs, New Jersey, 1975.

A good book on reports and business correspondence. Places emphasis on problems that arise in technical writing. Sections of the book are devoted to organization, style, and mechanics. Contains a selective handbook on language fundamentals.

Tichy, H. J., *Effective Writing for Engineers, Managers, Scientists*, John Wiley & Sons, New York, 1966.

Another book designed as a course of self-improvement in technical writing. Much good material on organization and style. Particularly useful discussions of getting started at writing and of supervisory review.

Ulman, Joseph N., Jr., and Jay R. Gould, *Technical Reporting*, third edition, Holt, Rinehart, and Winston, New York, 1972.

A popular book with technical writers and editors. Contains much material on writing and layout. Covers a variety of types of technical writing. Detailed presentation on style, grammar, punctuation, mechanics, and use of tables and illustrations. Excellent reference book for the report writer.

BASIC REFERENCES ON WRITING

Fowler, H. W., *A Dictionary of Modern English Usage*, second edition, revised by Sir Ernest Gowers, Oxford University Press, New York, 1965.

A classic work brought up to date. Deals with a broad range of questions that are of concern to the serious writer—questions of grammar, syntax, style, spelling, punctuation, and so on.

Perrin, Porter, *Writer's Guide and Index to English,* fifth edition, Scott Foresman, Chicago, 1972.

An excellent reference book for anyone who writes. Full discussions of and a useful index to English usage.

Strunk, William, Jr., and E. B. White, *The Elements of Style,* Macmillan, New York, 1959. (Paperback edition, 1962).

An effective presentation of the elements of good style. Short, concise, and direct.

JOURNALS AND SOCIETY PUBLICATIONS

Journal of Technical Writing and Communication.

A quarterly published by Baywood Publishing Company, 43 Central Drive, Farmingdale, New York 11735.

Technical Communication.

The quarterly journal of the Society for Technical Communication, 1010 Vermont Avenue, N.W., Washington, D.C. 20005.

The Technical Writing Teacher.

A journal of The Association of Teachers of Technical Writing, published three times each year. Send inquiries to John S. Harris, English Department, Brigham Young University, Provo, Utah 84602.

The Society for Technical Communication has other publications that can be obtained from its headquarters at the address shown for its journal. These include a useful booklet on the teaching of technical writing and a growing Anthology Series:

Thomas E. Pearsall, *Teaching Technical Writing, Methods For College English Teachers,* 1975.

Proposals . . . and Their Preparation, Anthology Series, No. 1, 1973.

H. Lee Shimberg, Ed., *Managing a Publications Department,* Anthology Series, No. 2, 1974.

Helen G. Caird, Ed., *Publications Cost Management,* Anthology Series, No. 3, 1975.

Lola M. Zook, Ed., *Technical Editing: Principles and Practices,* Anthology Series, No. 4, 1975.

James G. Shaw, Ed., *Teaching Technical Writing and Editing,* Anthology Series, No. 5, 1976.

Index

Abstracts and summaries, distinction between, 82
  examples of, 82-84
  importance of, 20, 82, 83
  for managers, 20, 23, 83
  and report form, 52, 53
  use of, 20, 23, 52, 53, 82-84
  writing of, 8, 20, 23, 82-84
Active verbs and passive verbs, 64-66
Alternative patterns of organization, examples of, 26, 32
  and orders for presenting ideas, 39-40
  seeking, 7-8, 30-31
  selecting, 31-33, 39-40
    criteria for, 7-8, 25-26, 33
Analysis of communication problem, 6-7, 10-24
Application of report design, 8, 59-79
Article report form, 52, 53
Audience, anticipating questions of, 19-20, 37
  background of, 23
  and decision making, 3-4, 18-20, 21-22, 25-26, 37, 83
  as element of communication problem, 7, 10, 17
  and illustrations, 42-44, 49
  informational needs of, *see* Informational needs
  reading habits of, managers, 19-20
    technical staff, 20-21
  and report form and layout, 16, 50, 52, 54
  and use of reports, 15-17
  view of writing, 40-42

"Big" words, 77-78

Chronological order, 3, 7, 39
Coherence, 70-72
Communication problem, analysis of, 6-7, 10-24
  and technical problem, confusion with, 2-3
    solution of, 29-30
Conciseness, 74-76, 79
Content of reports, in body and appendix, 37
  definition of, 26-29
  gathering, 29-30
  organizational and informative, 40-42
  organizing, 25-26, 27-29, 30-33, 35-36, 37-40
  selecting, 36-37
    for audience, 17-19, 20-21, 21-23
    for purpose, 10-15
Conveyor system example, 21-22, 25-26, 27, 35
Correspondence, characteristics of, 16, 50
  forms, 50, 51
  impersonal style and tone in, 65-66
  negative tone in, 67-69
  writer-reader relationship in, 16, 50
Correspondence report forms, 51-52
Criteria for overall organization, developing, 25-26
  using, 26, 31-33

Dangling modifiers, 73-74
Deadwood, 74-75, 76
Deductive approach, to defining content, 27
  to outlining, 27, 34-35
Density of style, 72-73, 76, 78-79
Design, in engineering, 6-7
  of illustrations, 45-46, 49
  of reports, 8, 34-57
  and writing process, 6-8
Design approach, to technical writing, 6-9

Effective Writing, 59
Elements of Style, The, 79
Emphasis, by illustration, 44
  of important material, 8
  by page layout, 55-57

Faulty sentence structure, 72, 73-74
Final copy, 62-63
First draft, getting started at, 59-60
  revising, 60-61, 62, 63
  and style and tone, 63, 70, 72, 77
  writing, 60-62, 63
Formal outline, and informal outline, 35
  need for, 35-36
Formal report form, 16, 50, 52, 53, 54, 55
Form report, 51
Form of reports, choice of, 16, 49-52
  as containers, 50, 51, 52
  as distinct from layout, 49
  and prescriptions, 49, 52
  types of, 16, 50-52, 53, 54, 55
  and writer-reader relationships, 49, 50, 52
Future Shock, 1

General-to-specific order, 39
Grammar, faulty, 73-74
  and style, 70, 72

Headings and subheadings, functions of, 8, 15, 16, 49, 52, 54
  as graphic representation of content and organization, 8, 16, 52, 54
  handling of, 52, 54-55, 56
  use, influence on, 15, 16
  see also Internal layout

Illustrations, designing and selecting, 45-46, 49
  and final copy, 63
  guiding reader's use of, 42-43, 49
  importance of, 42
  locating, 42, 46-47, 49
  in support of text, 42, 43-45, 49
Impersonal style and tone, *see* Style; Tone
Improvement curve in writing, implications of, 85-86
  nature of, 85
Inaccurate word choice, 77
Indirect sentence openings, 75-76
Inductive approach, to defining content, 27-29
  to outlining, 27-29, 34-35
Inflated tone, 69
Informal outline, 29, 35
Informational needs, of managers, 17-19, 21-22, 22-23
  of readers generally, 21-23
  of technical staff, 20-21
Informative content, 40
Internal layout, choice of, 49
  as distinct from form, 49
  functions of, 8, 15, 16, 49, 52, 54
  handling of, 52, 54-55, 56
  and prescriptions, 54, 55
  use, influence on, 15, 16
Introductions, faulty concepts of, 60
Investigation of alternatives, 7-8, 25-33

Jargon, 69, 79

Layout, *see* Internal layout; Page layout
Letter, 9, 16, 50, 51, 65-66
Letter report form, 51-52
Listing details, in prepositional phrases, 76
  in stacked modifiers, 78-79
Listing ideas, 70-71

McDaniel, H. C., 17
Main idea of a report, 35
Management audience, 3, 17-20, 21-22, 22-23, 39, 83
"Map" paragraphs, 40-42
Memo, 10-12, 16, 50, 51, 65-66
Memo report form, 51-52
Misplaced modifiers, 74
Modifiers, dangling, 73-74
  misplaced, 74
  stacked, 78-79

Negative tone, effects, on audience, 67-69
    on subject matter, 66-67

Orders for presenting ideas, 3, 39-40
Organization, of reports, alternative patterns of, 7-8, 25-26, 30-33, 39
    and internal layout, 8, 15, 16, 49, 52, 54
    revealing, 40-42
    selected for, audience, 3, 7, 8, 19-20, 26, 33, 37-39
      purpose, 10-15, 33, 37-38
      use, 15-16
Organizational content, 40-42
Outline, content, 34-35, 36-37
  deductive approach to, 27, 34-35
  detailed development of, 8, 34-37, 38-40
  formal and informal, 35-36
  as guide for writing, 29, 34, 36
  inductive approach to, 27-29, 34-35
  main idea, 35
  order, 39-40
  relationships, 38-39

Page layout, 49, 55-57
Pandora example, 31-33, 35
Paragraphs, listing of ideas in, 70-71
  and transitions, 71-72
Perrin, Porter G., 69
Personal style, 64, 65-66
Physical characteristics of reports, 16-17
Prepositional phrases, overuse of, 76
Primer style, 70-71
Pronouns, personal, 64, 65-66
  vague reference of, 73
Purpose of reports, classification by, 13-15
  in determining content and organization, 13-15
  effect of, examples, 10-13
  as element of communication problem, 6-7, 10

Quiller-Couch, Sir Arthur, 79

Readers, *see* Audience
Reader's view of written material, 40-42
Reading habits, of managers, 19-20
  of technical staff, 20-21
Report form, 50-52, 53, 54, 55
  article, 52, 53

correspondence, 51-52
form, 51
formal, 16, 50, 52, 53, 54, 55
see also Form of reports
Revision, 8, 60-61, 62, 63, 69-70
Rickover, Vice Admiral Hyman A., 79
Roundabout phrases, 75

Sanford, Bruce, 40
Scientific and technical writing, characteristics of, 1-4
difficulties of, 4-5
growth of, 1
quality of, 85-86
Sentences, faults in, 72, 73-74, 75-76
listing details in, 72-73, 76
short and simple, overuse of, 70-71
wordiness in, 72, 74-76
Sources of information, 29
Stacked modifiers, 78-79
Strunk, William, Jr., 79
Style, and audience, 23, 63, 64, 65, 66
definition of, 63
faults of, in paragraphs, 70-72
in sentence structure, 72-76
in word choice, 76-79
and grammar, 70, 72
impersonal, appropriate use of, 64-66
inappropriate use of, 64-66
nature and relationship of, 64
and personal style and tone, 64, 65-66
weaknesses of, 64-65
importance of, 63, 69
jargon, 69, 79
personal, 64, 65-66
and tone, 64-66, 68-69
in writing and revising, 62, 63, 69-70
Subordinate clauses, use of, 76
Summaries and abstracts, *see* Abstracts and summaries

Tables, 42, 45, 46
Technical problem, and communication problem, confusion of, 2-3
solution of, 29-30
Technical staff audience, 20-21
Tichy, Dr. Henrietta, 59
Toffler, Alvin, 1
Tone, and audience, 23, 63, 64, 65, 66, 68, 69
definition of, 63
impersonal, 64-66
importance of, 63-64
inflated, 69
negative, 66-69
personal, 64, 65-66
and style, 64-66, 68-69
in writing and revising, 63
Transitions, means of providing, 72
need for, 71-72
Types of reports, by form, 50-52, 53, 54, 55
by purpose, 13-15

Use of reports, as element of communication problem, 6-7, 8, 15
influence of, on form, 16, 23-24, 49-52
on internal layout, 16, 23-24, 54
on organization, 15, 23-24, 33
on physical characteristics, 16-17, 23-24

Westinghouse study, 17-20
White, E. B., 79
Word, choice of, 76-79
inaccurate, 77
in stacked modifiers, 78-79
unnecessarily "big," 77-78
Wordiness, 72, 74-76
Writer's Guide and Index to English, 69
Writer's view of written material, 40-42
Writing, activity of, 59, 60-63
faulty concepts of, 60
first draft, 59-60, 61-62
getting started at, 59-60
improvement curve, 85-86
process, features of, 8-9
stages in, 6-8, 24
for readers, 17, 21-23
and revising, 59, 60-61, 62, 63